TEN EQUATIONS TO
EXPLAIN THE MYSTERIES
OF MODERN ASTROPHYSICS

TEN EQUATIONS TO EXPLAIN THE MYSTERIES OF MODERN ASTROPHYSICS

FROM INFORMATION AND CHAOS THEORY TO GHOST PARTICLES AND GRAVITATIONAL WAVES

Santhosh Mathew

BrownWalker Press
Irvine • Boca Raton

BrownWalker Press / Universal Publishers, Inc.
Irvine • Boca Raton
USA • 2019
www.brownwalkerPress.com

978-1-6273-4720-4 (pbk.)
978-1-6273-4721-1 (ebk.)

Typeset by Medlar Publishing Solutions Pvt Ltd, India
Cover design by Ivan Popov

Publisher's Cataloging-in-Publication Data

Names: Mathew, Santhosh, author.
Title: Ten equations to explain the mysteries of modern astrophysics: from information and chaos theory to ghost particles and gravitational waves / Santhosh Mathew.
Description: Irvine, CA : BrownWalker, 2019. | Includes bibliographical references.
Identifiers: LCCN 2019938968 | ISBN 978-1-62734-720-4 (paperback) | ISBN 978-1-62734-721-1 (ebook)
Subjects: LCSH: Astrophysics--Popular works. | Astronomy--Popular works. | Physics--Popular works. | Mathematics--Popular works. | Entropy. | Outer space--Exploration. | BISAC: SCIENCE / Physics / Astrophysics. | SCIENCE / Astronomy. | SCIENCE / Physics / Mathematical & Computational.
Classification: LCC QB461.3 .M38 2019 (print) | LCC QB461.3 (ebook) | DDC 523.01--dc23.

Table of contents

Foreword

What if the reality that we see before us is only temporary? Or, perhaps it is not what we think it is. Where then do we look to grasp the essence of human existence? What if the concept of eternity is illusory? How then do we make sense of human existence? These are just a handful of the questions that Santhosh Mathew asks his readers to contemplate. These are not easy questions to answer, and when reading his book, it becomes clear that the answers to these questions are not likely to be found in many of the places where humans have typically searched for truth or have tried to garner deeper understanding. To further complicate our arrival at the truth, the noise and abundance of data that contemporary society drops on our lives diverts our ability to contemplate and understand our existence, and the world around us, with great clarity. Within Mathew's book, equations are the antidote that is suggested for the forces that stymy our understanding.

The questions posed above can be unsettling; however, Mathew acknowledges and welcomes the astonishment that accompanies the enduring quest among humans to find meaning.

As Mathew points out, humans strive to construct order from chaos in their lives through borders, systems, and tangible structures, yet all of these human ecosystems are fleeting since humans, themselves, are transient. This book, in focusing on the timeless qualities of equations, conveys both the joy of discovery and the possibilities that arise when one honors questions over answers, process over outcomes, and journeys over conclusions.

In order to thoughtfully explore such enduring questions, and to unravel select truths, Mathew suggests the importance of equations and shows readers that, due to their continual lifespans, equations offer us a link from the past to the present and into the future. Within his narrative, Mathew refers to equations as elegant, "timeless and spaceless" and can be, in the right hands, and contemplated through the right minds, a "beautiful painting hiding a message." These descriptions present equations as invitations to connect moments of our history and discern the elements that underlie our existence. Through equations, we find not only a welcoming invitation but also deeply embedded meaning that has been formed and reformed throughout history, but not always well understood, and, at times, highly misunderstood.

While the book is structured to enlighten readers on ten questions, the prevailing themes are far from linear. Mathew weaves the present with the past; science with philosophy; and the spiritual teachings of religion with the discoveries of science. To further synthesize his themes, Mathew draws upon a wide range of writers, philosophers, scientists, and sources, including Herman Hesse, Carl Sagan, Franz Kafka, Stephen Hawking, Fritjof Capra, and the Bhagavad Gita.

Mathew takes a deep dive into each of the ten equations and illuminates how all of them hold basic truths. He artfully draws our attention to how equations connect to human behavior in the past, such as cave drawings by early man, to today's newscasts that construct narratives from disparate sources and define and redefine our contemporary social worlds. Other equations within the book's chapters demonstrate how everyday concepts, such as, what goes up, must come down, are used to enlighten our understanding of our wider world, sometimes with validity, sometimes with incomplete understanding, and sometimes even leading to popular misnomers.

At the heart of Mathew's book is the concept of entropy, derived from the Greek word *entrope*, meaning to turn toward or transform. While humans crave certainty, and often seek stability and absolute truths, Mathew shows his readers that far more can be learned through disorder, decay and chaos. Through entropy, a more nuanced truth and higher meaning emerge than could ever be realized through an existence predicated on certainty. Rather, through the lens that Mathew uses, probability offers far more potential than determinism. In many ways, Mathew's book invites its readers to become blissfully lost in novel findings, unexpected directions, and conclusions that guide explorers to a boundless array of unimagined pathways.

While Mathew presents equations as gateways to human understanding, the book illustrates that, throughout history, many humans have sought absolute truth, and even a sense of justice, through philosophical and spiritual texts. While Mathew does not dismiss the quest to understand life through a wide range of sources, this book shows how these sources simply illuminate and act as metaphors for the discoveries of science rather than supplanting science or serving in any way as equivalents.

Mathew's central message is that science is a journey rather than a destination. While humans may look at other sources and crave singular explanations, or wish to place their trust in ancient texts, it becomes clear that these sources have limitations, at their best, and misguided notions at their worst, especially when used to support nationalism or other aggressive agendas. Mathew's book shows that human existence is dependent on continual discovery that only deepens and expands over time. Through this view, some of the worst enemies of the human race may very well be premature conclusions and easily arrived upon destinations since they may stifle the human quest for meaning. In essence, at the core of this book is the alluring and insatiable notion of curiosity. Through this lens, nothing is settled; therefore, everything remains possible.

Debra Leahy, Ed.D.
Chief Academic Officer, Boston, MA.

Preface

We all know transience is a part of life, but it is not just part of our life. Everything is transient, from the fundamental particles that make up every living thing to the ever-expanding universe—on a different time scale, though. We latch on to many things and ideas and assume they will remain forever, but eternity, whether on Earth or in heaven as some would like to believe, in a broader sense, is an illusion.

In his final novel, *The Glass Bead Game* for which he won the Nobel Prize for Literature in 1946, Hermann Hesse wrote:

> No permanence is ours; we are a wave
> That flows to fit whatever form it finds:
> Through night or day, cathedral or the cave
> We pass forever, craving form that binds.

The inspiration to present this work emerged from our transient human nature and the universe we inhabit. At the same time, we have a deep desire to discover the laws governing everything around us, which I consider eternal. Our ancestors gazed in wonder at the world around them, just as we do today. They thought of the big questions, as we do even now. How did the universe begin? What is it made of? Where do human beings fit in this great cosmic scheme? Finally, is there a meaning to it all?

The meek and the powerful lived on this planet; so did the rich and the poor. Kings and emperors ruled the Earth, but they have all vanished and become part of the grand universe. What did they leave behind? We know that mighty structures can crumble without a trace, that species can become extinct and even their fossils may not be found in the long run, and in the far future, our planet may be gone. So, what will remain after we disappear? Definitely not any of the material objects we have created. Today's strongest structures can become fundamental particles of nature in the far future. We call this process natural decay. Yet, we can preserve something: the secrets we have unveiled about nature using our most rational tool called science. The most elegant way we can keep this information is in the form of equations.

The equations represent our intellectual ability to explore and find the irrefutable laws of nature and our insatiable curiosity to know more. We discover universal laws from patterns of nature we observe or data we collect, and this, in my view, is the greatest achievement of humankind. This human talent supersedes any other accomplishments, and so these equations represent an eternal state—they are timeless and spaceless.

No doubt humans will leave a lasting impression on the Earth's surface, or we may terraform other planets. We will play with mass and energy and create one form or the other, or we will travel to other parts of our galaxy on light sails. We may edit the genome to design a perfect species or send radio waves millions of light-years away, but none of these will stand the ultimate test of time. Nature will be able to wipe out all signs of human existence on Earth or any other place we occupy.

As humans, we embrace specific beliefs and often imprison ourselves within the culture that we create and are proud of. We create borders and language, all with the hope of having some control over our surrounding world, and then we realize the inevitable futility of those controls. We need something in which to trust permanently, and we call it eternal. It is unimaginable for us to spend our lives in a random universe without any specific purpose and to leave things beyond our control. That will bring only despair.

I believe equations are the only way to know the universe even if we are not close to presenting a definite answer to many perplexing questions. The equations are not mere symbols with abstract meaning; rather they are like beautiful pieces of artwork that capture the essence of nature. Like a beautiful painting hiding a message, the equations unveil the mysteries of our nature and the process of knowing. That is the most wonderful experience! These equations represent absolute laws of nature, and they do not provide us with many choices. We must simply obey.

Despite all the fascination surrounding these equations, I must caution you that nature and the laws of nature are not fine-tuned for humans. We are a transient species that happened to find ourselves here on this tiny planet about a quarter million years ago. As Carl Sagan noted, "The universe is not required to be in perfect harmony with human ambition." Yet, our voyage continues.

In the coming chapters, you will see ten equations that will outlast even the most enduring signs of our civilization.

These equations have changed the way we live and view the world and will continue to do so. They have the potential to take us from planet to planet and perhaps to make us a cosmic species or to destroy the last strand of DNA. I hope that you enjoy the experience of knowing these ten equations closely (some are well known mathematically) and that you develop an appreciation for them. They are intangible but can create a tangible world; nevertheless, they remain truly eternal.

Acknowledgements

T
hroughout the process of writing this book, I received help from quite a number of people. This work could not have been accomplished without the help of these individuals. A great debt of gratitude is due to everyone who supported me during the course of this work.

My deepest gratitude and thanks to Jeffrey Young, editor, Universal Publishers/BrownWalker Press, who relentlessly encouraged me to complete this project. He oversaw every aspect of the publication of this work, which in fact began the day I sent out the book proposal. The prompt responses to all my queries and the valuable advice I received from him enabled me to keep to the schedule.

I would like to express my sincere thanks and gratitude to Debra Leahy, Chief Academic Officer, New England College of Business, Boston, who enthusiastically took the time to read the entire manuscript and wrote a detailed foreword to accompany this book. I am deeply indebted to her for the unyielding support and encouragement I received.

I wish to express my deep sense of gratitude to Professor Abraham Loeb, Chair of the Astronomy Department at Harvard University, who graciously agreed to comment on my book. Professor Loeb has always been a true source of inspiration and support for me and dedicated his time and expertise to meet all my requests on several occasions.

My sincere appreciation goes to Ishwar K. Puri, Dean and Professor of Engineering at McMaster University, Canada, for commenting on this work. Needless to say, Professor Puri has always offered tremendous support and expressed interest in my research since our first meeting about fifteen years ago.

I would like to thank my wife, Sumy, and son, Nathan, who have given me encouragement and support over the years. Furthermore, I wish to express my gratitude to my parents and siblings who have always supported and encouraged me in this endeavor.

Finally, I want to thank all those who provided support, talked things over, read, wrote, offered comments, and assisted in the editing, proofreading, and design of this book.

Santhosh Mathew

Introduction:
the origin of equations

I t has been said that humans, and all that comes of humans, fades, and so do any living things that spring from them. Nevertheless, if there were something that defied this rule, it would be those equations that humans invented and that serve as a tremendous resource to comprehend and appreciate the complex universe in which we live. Moreover, those equations have the potential to create a new world, maintain our own world, or destroy the world. They are visible everywhere—from battlefields to graveyards, from the heart of a flower to the core of a galaxy. We, the transient humans, continue to search for these eternal equations, and they provide a timeless experience in this ever-changing universe. There is no time and space in the world of equations as they transcend even the boundaries set by human imagination. Moreover, they inspire us to embark on a journey to comprehend the magnificent universe and its workings.

This book introduces ten equations that unravel the mysteries of our universe, and I hope it will take readers on a journey of self-discovery where they will learn history, science, and the significance of these equations in their lives in addition to the mathematical beauty of these equations.

What is so remarkable about equations? They have a universal beauty attached to them that transcends earthly borders. Moreover, they do not have an aspect that we associate with other things, (e.g., cultures, geographic locations, or nationalities). The equations are truly universal. It is no wonder Galileo said, "Mathematics is the language in which God has written the universe." Needless to say, it is quite natural to use mathematics to express the laws of nature in an elegant way.

Look at any product or social construct we deal with. All carry baggage associated with nationality and often relish those attachments, but the equations transcend those man-made systems. Using pictures and symbols, the ancient Babylonians and Greeks tried to explain relationships between different things. These bits and pieces set the stage for equations to appear in the human world. Essentially, the idea behind equations was the desire to connect seemingly different things with a common thread. Later on, these threads were woven into the fabric of the modern world that we live in—from electricity to the internet, from atoms to artificial intelligence, from cannonballs to rockets, and the list goes on.

This idea of interconnectedness and exploration is so fundamental to human existence that it is unlikely to vanish from the human psyche. In fact, today's world of networked systems exhibits another example of this deep-rooted desire to be connected. Long before any modern systems were even imagined, our ancestors displayed the same intention as we do now.

The equations come in myriad forms and are packed with symbols and notations that seem abstract, but they enrich our daily lives in various ways as you will see in the coming chapters that introduce ten different equations, which are also tied to various astrophysical phenomena. I have to admit that, even when some of these equations seem quite familiar, we do not fully understand their depth, although we use them in various forms—a humbling reason why we should continue to learn more about equations that connect everything in our universe.

The current form of equations, identified by the equals sign (=), appeared for the first time in the sixteenth century. That symbol, now universally accepted by mathematics as the symbol for equality, was first recorded by Welsh mathematician Robert Recorde in *The Whetstone of Witte* in 1557. In his book, Recorde explained his design of the "Gemowe lines" (meaning *twin* lines, from the Latin *gemellus*). The first-ever equation with the now common equals sign, $14x+15=71$ in modern terms, was mentioned in his book. With the publication of this book, Recorde is credited with introducing algebra with a systematic notation to England.

The equations are not merely symbols and variables, as they seem to be. In fact, they combine everything we know about our universe and provide a deeper understanding about our natural world. They are the ultimate manifestation of human ingenuity rooted in the laws of nature. In light of new evidence, we will abandon old ideas and theories and humans will depart the Earth, but the equations will remain eternal.

CHAPTER 1

The equation that gave us a digital life

No theory of physics that deals only with physics will ever
explain physics.
—*John Wheeler, The Intellectual Digest (1973)*

Abstract

*On a fundamental level, what is our universe made of? Space, time,
matter, and energy—those are all familiar explanations. However, some physicists propose that information is the fundamental
ingredient of our universe, not force fields or space-time as generally
accepted now. Therefore, if information is so fundamental and essential to this universe, the equation that launched the information
revolution must be considered, not only in communications, but also
in the fundamental understanding of this universe. The idea of the
holographic principle makes more sense if we can link it to information theory. This chapter explores the significance of the Shannon*

equation in the context of the holographic principle, a radical idea in modern astrophysics.

The idea that the information exchange among physical processes could be the foundation of this universe, though it sounds strange, is not very hard to understand. Imagine a computer, which is essentially an information-processing machine. However, without the software to run it, it is merely pieces of metal and glass. Alternatively, think about the actions that should happen in our body cells. We cannot even picture them without the information from DNA. So, we could say with a greater degree of confidence that, in the absence of information, objects are not what they seem to us. It is such simple logic.

Similarly, the universe that we live in and observe could be considered as a huge system operating with information as the underpinning factor. Proponents of the holographic principle say that we should explore this aspect to know the fundamental truth. Some theorists argue that to marry quantum mechanics with gravity requires this radical approach that put information at the core of this universe and that information is much more significant than space-time or matter and energy.

Generally, physics describes the fabric of our universe as space-time where the interplay between matter and energy embodies everything in this universe. In addition, the general theory of relativity says that, at its most basic level, this fabric should be smooth and continuous. Yet, if we examine the fabric of space-time on a fundamental level, it is not continuous, as we might imagine, but simply grainy. These grains act like dots in an image providing us with the vision of a three-dimensional universe. This idea led to the so-called holographic principle proposed by Nobel-Prize-winning Dutch theoretical physicist Gerard 't Hooft (1993). The evidence to support this

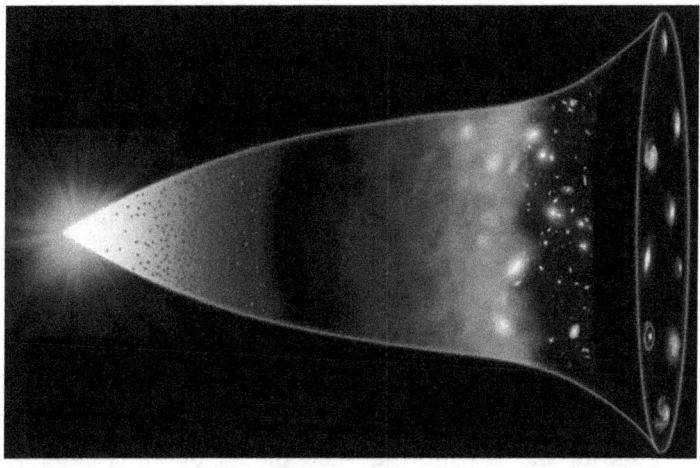

Fig. 1.1. *The universe is shaped like a giant bell lying on its side. The bell was struck nearly 14 billion years ago and emerged in oscillating waves from the singularity. Is our universe simply made of information? Image courtesy of the Perimeter Institute, Waterloo, Ontario, Canada.*

principle came from the irregularities in the cosmic microwave background, popularly known as the afterglow of the Big Bang. According to the holographic principle, our world is simply an image of the information stored on a two-dimensional projection similar to a holographic image (Fig. 1.1). The holographic principle tells us that when we look at this two-dimensional projection the right way, it gives us the view of a three-dimensional universe. In simple terms, our 3D world is a mere gift from the underlying information. Although not every physicist buys into this argument, it received considerable attention from the physics community.

In the 1980s, even before this holographic picture of our universe was presented by researchers, the legendary physicist, John Wheeler, hinted at the significance of information in understanding our cosmos. He described his evolution in physics as falling into three periods, thereby signaling the deep connection between quantum mechanics and information theory. "I think of my lifetime in physics as divided into three periods. In the first period... I was in the grip of the idea that Everything Is Particles... I call my second period Everything Is Fields... Now I am in the grip of a new vision, that Everything Is Information," that is, "It from Bit" (Wheeler, 1990).

Does this mean the fundamental reality is information? Not necessarily, and many scientists totally disagree. However, we need to recognize the fact that the descriptions of reality and information influence each other, and both are needed to account for and to paint the whole picture of our cosmos. To support his idea, Wheeler argued that an electron behaves like a particle or a wave depending on how we probe it. Information theory, similarly, postulates that all messages can be reduced to a sequence of "binary units," or bits, which are answers to yes or no questions. Essentially, Wheeler, in his 1989 essay, took the position that anything physical (it) derives its existence from information (bit) popularized by the phrase "It from Bit." So, if information is the root cause of the existence of physical objects, we can postulate theories that describe the universe in the language of information theory.

What is information theory?

In June 2016, the United Nations Human Rights Council passed a resolution making internet access a basic human right very much like the right to safe drinking water, sanitation, shelter, and basic services. The information interchange, like transportation, is so essential to the modern world that nations would be paralyzed and would collapse in the absence of the internet.

This relatively new phenomenon, a global network or internet that unfolded before our eyes, is essentially an exchange of information. It was not long ago, as many remember, that snail mail was the primary communication mode, and telegraph and telephone were not accessible to most people in the world. It has been estimated that currently more than four billion people use the internet worldwide (World Internet Users Statistics and 2018 World Population Stats, n.d.). For Generation Alpha, it would be practically impossible to imagine a time without the internet, just as it would be impossible for the previous generation to imagine a time without electric lights.

So, what caused this revolution right in front of our eyes? When did this new information age begin? Surprisingly, it is Claude Shannon's equation that really changed the way in which information, or bits, was effectively transmitted across the various systems. In fact, Shannon was addressing the fundamental problem of communication that involves the mechanism to reproduce a message exactly, or approximately, at one point that was selected at another point.

This equation was first published by Shannon as "A Mathematical Theory of Communication" in the July 1948 issue of *The Bell System Technical Journal.* In this fundamental work, he used the tools of probability theory and developed information entropy as a measure for the uncertainty in a message.

$$H = -\sum p(x) \log p(x)$$

H – Shannon entropy or information entropy
p – probability

The entropy of a message is related to the sum of the logarithm of the probabilities of each bit taking on a particular value. The

Shannon entropy equation provides a way to estimate the average minimum number of bits needed to encode a string of symbols, based on the frequency of the symbols. The physical meaning of information entropy is the absolute minimum number of storage "bits" needed to capture the information. Although Boltzmann (more about Boltzmann's equation in chapter 5) was the first person to formulate the logarithmic function to connect the average uncertainty with the probability of random variable, Shannon

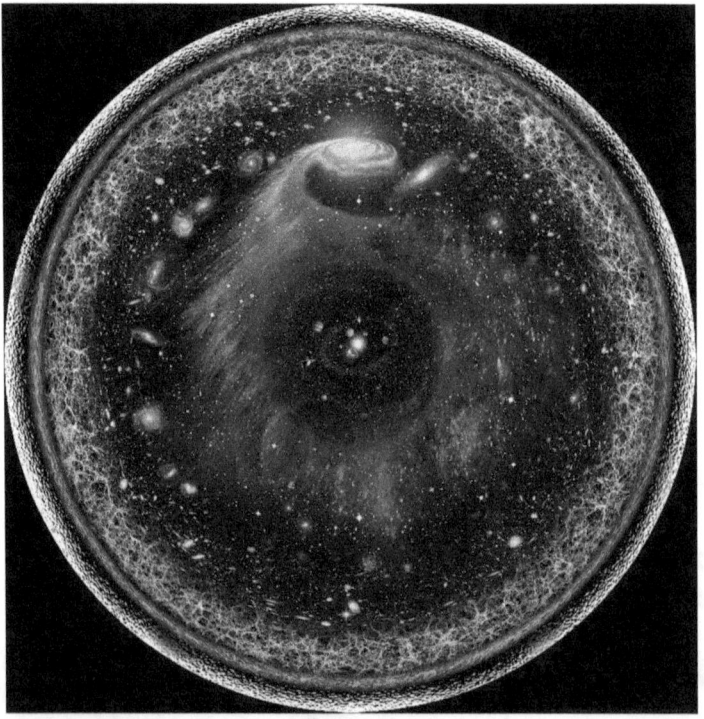

Fig. 1.2. *Artist's logarithmic scale conception of the observable universe with the solar system at the center. Can we create a universe out of information? Image courtesy of Pablo Carlos Budassi, Wikimedia Commons.*

extended this result to the communication scenario to propose two theorems—source-coding theorem and channel-coding theorem—that are the basis of modern communication technology.

We know that the earliest form of human communication was drawing pictures or creating patterns using available resources. From that, primitive stage humans evolved to encode information using zeros and ones. Now, here is an equation that tells you information is "additive"—that is, that the total information contained in two, three, four, or a billion unrelated events is equal to the sum of the information in each one (Becker, 2015). So, theoretically, if we have the means to do so, we can record all information about this universe in bits and possibly reconstruct it (Fig. 1.2). However, practically, finite resources and restrictions imposed by the laws of physics prohibit us from doing so. Therefore, we won't be able to create a cosmic computer now to run the actions of this universe, yet some real-world experiences inspire us to capture the meaning of it from bit as discussed below.

Can bit generate it?

Normally, it's hard to conceive how bit (information) alone can generate it (physical stuff) as the legendary physicist John Archibald Wheeler summarized in his 1989 essay with the catchphrase, "it from bit." However, when I think back to my childhood experiences, long before I came to know information theory or the holographic principle, one of those experiences makes much more sense in light of the phrase "it from bit."

Throughout my early school years, in a south Indian village, I remember spending a great deal of time outside the walls of my home interacting with the natural environment—obviously

I didn't have many choices. In those days, my native village had the look of a flatland that displayed a romance of many dimensions. A huge rock, seemingly indistinguishable from an asteroid belt object, occupied the heart of the village, and I always thought this stone had been transplanted from the remote corners of the solar system by some astronomical process. Every object and every piece of information was conserved in the village by the laws of nature. People believed there were no absolute realities but only descriptions of realities that existed without any knowledge of quantum mechanics or even any philosophical views that I was aware of.

The elegant movement and perfect rhythmic life cycle of people and animals obviated the need for mechanical timepieces; though a couple of villagers displayed watches on their wrists, but they never looked at them. The relative nature of time and space was fully comprehended by everyone. The village maintained a constant ratio of births and deaths, and everyone knew everything about others. Complete transparency was achieved without electronic data or documents. Sex was a taboo, yet it was performed only to maintain the population and, understandably, never discussed even between partners.

Information came to the village in the form of a local newspaper printed in a faraway town and delivered by bikes that followed rudimentary Newtonian mechanics. The other information system involved radio waves that reached the village intermittently as the countryside hills surrounding the village offered a barrier for the electromagnetic waves. Paradoxes were plentiful in the village, for example, laws were followed without laws. And when Yama (the God of death in Hindu mythology) came as a cosmic equalizer, the village people acknowledged his visit by either ringing church bells loudly, but in simple harmonic motion, or by diffusing smells of Pooja[1] materials from their homes. Unaware of any of these rituals, Yama visited the village, taking away people and animals on a regular basis. No one complained as the acts were in compliance with the cosmic law.

In 1979, the news of Skylab's apparent fall reached the village. For the next two months, the social and religious events in the village were filled with narratives of the potential destruction that could be wrought by the falling space station. The Hindu temple performed Pooja and quoted relevant texts from Ramayana and Mahabharata (ancient Hindu texts) describing how their gods and the evil-spirited asuras (demons) had floated such objects in the past. The pastor invoked biblical quotes and warned this was a punishment from God for humans venturing into stars and skies, yet he was a bit soft on the United States that owned and operated Skylab. The few leftists in the village were apparently overjoyed by the fact that an American imperialist machine was coming down and used the occasion to celebrate the bright future of the Soviet Union.

As a ten-year-old boy, my curiosity regarding Skylab was more powerful than the death it could bring. So, every night I looked at the sky with fear in my heart, but my inquisitiveness destroyed the fear. Everyone owned a subjective description of Skylab, including me, and really wanted to see the amazing machine that was falling from the sky.

A few days later, on July 11, 1979, a churchgoer brought the fall of Skylab to the attention of the collective social psyche of the village. The newspapers had not been delivered and radios had not been working for several days, creating a perfect opportunity for everyone to chime in. The Christian

---

[1]Ritual performed by Hindus as part of worship.

devotee described seeing the night sky littered with heavenly lights and the machine falling through the sky like a star as described in the Bible. He even claimed to have seen biblical verses from the Book of Revelations inscribed on the fireball. According to the Hindu priest's account, Skylab resembled the chariot of Arjuna piloted by Lord Krishna (a scene that looked a lot like the epic battle of Mahabharata) crushing everything on its path. These narrations were widely accepted and, by now, everyone had created an image of Skylab in his or her mind.

The following weekend my elder brother brought home some magazines from his college town, and I curiously scanned the pages. One page had a fussy picture of Skylab as it fell into part of the southeastern Indian Ocean near western Australia, about 3,000 miles away from our village. In fact, no one in the village was able to see Skylab or its apparent fall as described above, yet everyone saw it.

After several years, I realized that Newtonian physics was sufficient to explain the fall of Skylab. Yet, I wasn't sure how or why everyone had created the image of an object (it) from whatever information (bit) he or she had. When I read Wheeler's essay that explained it from bit, it made perfect sense, and I understood why everyone had seen Skylab. This radical notion is even more relevant in the contemporary era where narrative (bit) construction of reality (it) is a norm. Furthermore, I am absolutely convinced it is the same explanation I need in our contemporary political age where narratives create facts and alternate facts. Once again, it from bit.

So, if we live in an information universe and Shannon's equation provides the most effective way to transmit information, we can imagine an eternal state in which all information is recorded in bits and stored or transmitted, though in a purely theoretical sense. For now, entertaining such a thought would be a pure imagination of mind, yet I feel we must explore the power of this equation to gain some sort of eternal existence. Paradoxically, this equation rests among the dead as my journey to Mount Auburn Cemetery (Massachusetts) revealed.

The equation for eternity rests among the dead

If you mention that you are visiting a cemetery, folks will show their sympathy by asking caring questions about your lost loved ones. However, if you want to know how to achieve eternity by encoding all your information onto digital bits, you should begin by looking at an equation that rests among the dead.

In the Mount Auburn Cemetery (Massachusetts), among the gravestones of many revered men, lies the unique grave of Claude Elwood Shannon. The back of the tombstone displays Shannon's entropy formula (Figs. 1.3 and 1.4) that launched the digital information age in the 1950s. All the information that pops up in a single Google search, or compressing huge files to a few binary digits, is digitally possible because of Shannon's equation. The equation provided a way to transmit information in the form of bits as we do now. As scientists engage in finding ways to satisfy the demand to "live forever" by exploring possibilities in cryogenics or biotechnology, the future will most likely require us to find a solution for eternal existence in digital clouds. If so, Shannon's entropy equation will be the key—whether information is stored in bits or qubits.

Fig. 1.3. *Shannon's entropy equation etched on his gravestone in Mount Auburn Cemetry, MA. Photo courtesy of the author.*

The Shannon equation, as discussed earlier in "What is information theory?", conveys one of the most powerful messages that physicists are exploring (i.e., that information is the fundamental building block of the universe, not space and time or matter and energy as we believe). This whole idea that everything is made up of

Fig. 1.4. *Shannon's gravestone in Mount Auburn Cemetery, Middlesex County, Massachusetts. Photo courtesy of the author.*

information even prompted some philosophers (Bostrom, 2003) to propose the idea that we almost certainly are living in a computer simulation. It follows that there is a significant chance that we will one day become post-humans who run ancestor-simulations. However, this is false unless we currently are living in a simulation.

The simulation argument has generated deeper debates among physicists and philosophers. If such notions were to be entertained, Shannon's formula would offer a solution to overcome human mortality that would enable us to generate and transmit every bit of information that makes us. Perhaps the same equation that created the digital information highway would help us to accomplish that goal in the future. In other words, eternity would be in digital form, not biological, as we tend to think. If information were the root cause of origin (universe and everything else that followed), then one could argue that the solution for eternal existence would be the same.

Finally, Fermat's principle, or the principle of least time, named after French mathematician Pierre de Fermat, says that the path taken between two points by a ray of light is the path that can be traversed in the least time. Some physicists extend this argument to support the holographic principle as they speculate it's a manifestation of the economical character of nature. So, the information theory, especially Shannon's equation, provides a way to describe the universe with a minimum amount of information (Correa-Borbonet, 2004).

Regrettably, we still struggle with these theories and so do the scientists. Unfortunately, we have a long time to wait for eternity. Perhaps our notion of eternity as the permanent state of existence may need to be rewritten. Eternity, according to Hermann Hesse, is a mere moment, just long enough for a joke. Either way, our quest for eternity may not be realized any time soon, but Shannon's equation remains eternal.

Bibliography

Becker, K. (2014, April). Is information fundamental? Retrieved from http://www.pbs.org/wgbh/nova/blogs/physics/2014/04/is-information-fundamental/

Bostrom, N. (2003). Are we living in a computer simulation? *The Philosophical Quarterly, 53*(211), 243–55.

Correa-Borbonet, A. L. (2004, January 18). Holography and the Shannon's first theorem. Retrieved from http://adsabs.harvard.edu/abs/2004hep.th.... 1118A

Eddington, A. S. (1987). *Space, time and gravitation: An outline of the general relativity theory*. Cambridge: Cambridge University Press.

Hesse, H. (2013). *The glass bead game*. United States: Stellar Classics.

Hesse, H., & Creighton, B. (1929). *Steppenwolf*. New York: Holt.

Kenyon, I. R. (2011). *The light fantastic: A modern introduction to classical and quantum optics*. Oxford: Oxford University Press.

Pauli, W. (2000). *Theory of relativity*. New York: Dover Publications.

Recorde, R. (1969). *The whetstone of witte: London 1557*. Amsterdam: Theatrum Orbis Terrarum. New York: Da Capo Press.

Roman, S. (1992). *Coding and information theory*. New York: Springer.

Shannon, C. E. (1948). A mathematical theory of communication. *Bell System Technical Journal, 27*(3), 379–423. doi. org/10.1002/j.15387305.1948.tb01338.x

't Hooft, G. (1993). "Dimensional Reduction in Quantum Gravity," Retrieved from https://arxiv.org/abs/gr-qc/9310026

Welcome to the United Nations. It's your world. (n.d.). Retrieved from http://www.un.org/

Wheeler, J. A. (1989). Information, physics, quantum: The search for links. In *Proceedings III International Symposium on Foundations of Quantum Mechanics* (pp. 354–358). Tokyo.

Wheeler, J. A., in Florence Helitzer's "The Princeton Galaxy", *Intellectual Digest 3*, No. 10 (June 1973).

Wheeler, J. A., & Ford, K. (1998). *Geons, black holes and quantum foam: A life in physics*. Norton: New York, NY.

World Internet Users Statistics and 2018 World Population Stats. (n.d.). Retrieved from https://www.internetworldstats.com/stats.htm

CHAPTER 2

The equation that predicted ghost particles

I have done a terrible thing: I have postulated a particle
that cannot be detected.
—*Wolfgang Pauli as quoted in Physics and Beyond (1971)*

Abstract

Neutrinos are the second most abundant particles in the universe, after photons. They are often called ghost particles since they barely interact with anything. During the 1920s, scientists came to the conclusion that matter was built of only two kinds of elementary particles:, electrons and protons. Later on, neutrons were added to the list. Yet, scientists realized something was missing in the process of β-decay. In 1933, Enrico Fermi formulated his β-decay model by introducing Wolfgang Pauli's invisible particles named neutrinos. The discovery of neutrinos in 1956 inaugurated a new phase in astronomy and physics, and these particles continue to mesmerize us. Some researchers think neutrinos could answer the questions of existence, for example, "How is existence possible?"

Why are ghost particles so cool?

The world is quite familiar with the God particle (Higgs boson), the long-sought particle that was discovered in 2012. Although the existence of Higgs bosons was hypothesized in 1964, they became part of popular culture in 1993 when the physicist Leon Lederman coined the term "the God particle" in his popular science book, *The God particle: If the universe is the answer, what is the question?* Higgs Bosons were considered an important missing piece of the standard model that physicists use to describe elementary particles and their interactions. It provided an answer to a fundamental question: why objects in our universe have mass. That explains why galaxies, stars, planets, and any material objects have the form they hold. However, well before the God particle came into picture, the existence of neutrinos, popularly known as ghost particles, was revealed, and the equation that led to the prediction and eventual discovery of these particles is surprisingly simple.

While the God particle provides an explanation for *why anything has mass*, the ghost particles can potentially explain *why anything exists*. In 1930, physicist Wolfgang Pauli first predicted the neutrino in order to account for the apparent loss of energy and momentum that he observed when studying radioactive beta decays. However, Pauli himself was not confident about the detection of the particles as they were more elusive than any other known particles had been, and that made even Pauli skeptical about the existence of such a particle.

In fact, the existence of neutrinos was proposed as a solution to save one of the most fundamental laws in physics, the law of conservation of momentum and, specifically, the law of conservation of angular momentum associated with spinning objects. According to this law, the total momentum of the interacting particles before and after a reaction should remain the same. Consider the following equation formulated by Enrico Fermi to explain β-decay in which a neutron decays into a proton, an electron, and an electron antineutrino:

$$n \rightarrow p + e^- + \nu$$

n – neutron

p – proton

e – electron

ν – antineutrino

The law of momentum conservation requires the neutrino to be present as a particle in the above equation. In the absence of a neutrino, the total momentum before and after the decay of the neutron as shown in the equation would be different. This would lead to the violation of the law of conservation of momentum, something that is impossible for physicists to cope with (along with the law of conservation of energy). Thus, the existence of Pauli's "invisible" particle was accepted, and the name neutrino (small neutron) was coined by Enrico Fermi.

Protons and neutrons consist of fundamental particles called quarks. In the beta decay process, a neutron (made of one up quark and two down quarks) can transform into a proton (made of two up quarks and one down quark). This reaction can happen in a neutron within an atom or a free-floating neutron.

$$\underset{\text{(udd)}}{n} \rightarrow \underset{\text{(udu)}}{p} + e + \bar{\nu}e$$

When a neutron turns into a proton, a down quark within the neutron transforms into an up quark, changing the neutron into a proton (and changing the atomic element as a result). The momentum and energy of particles must be conserved in the process and, to account for that, an electron and an electron antineutrino must be released in the process.

Neutrinos come from a variety of sources, making them the second most abundant particles (just behind photons, the particles of light). They are produced during the birth, collision, and death of stars, particularly the explosions of supernovae, and they carry invaluable information about these sources. It has been assumed that the neutrinos that originated in the early universe are still travelling and passing through all material objects without any interaction. In our Sun, neutrinos are created and pass through our planet and other materials, including the human body. The following diagrams show (Figs. 2.1 and 2.2) the proton-proton fusion reaction in the Sun and the resulting production of neutrinos.

Beta Decay of a Neutron

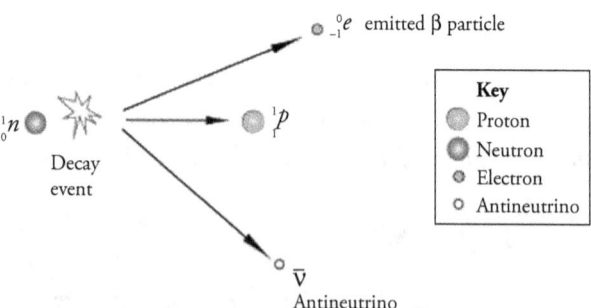

Fig. 2.1. *A neutron decays into a proton, electron, and an antineutrino.*

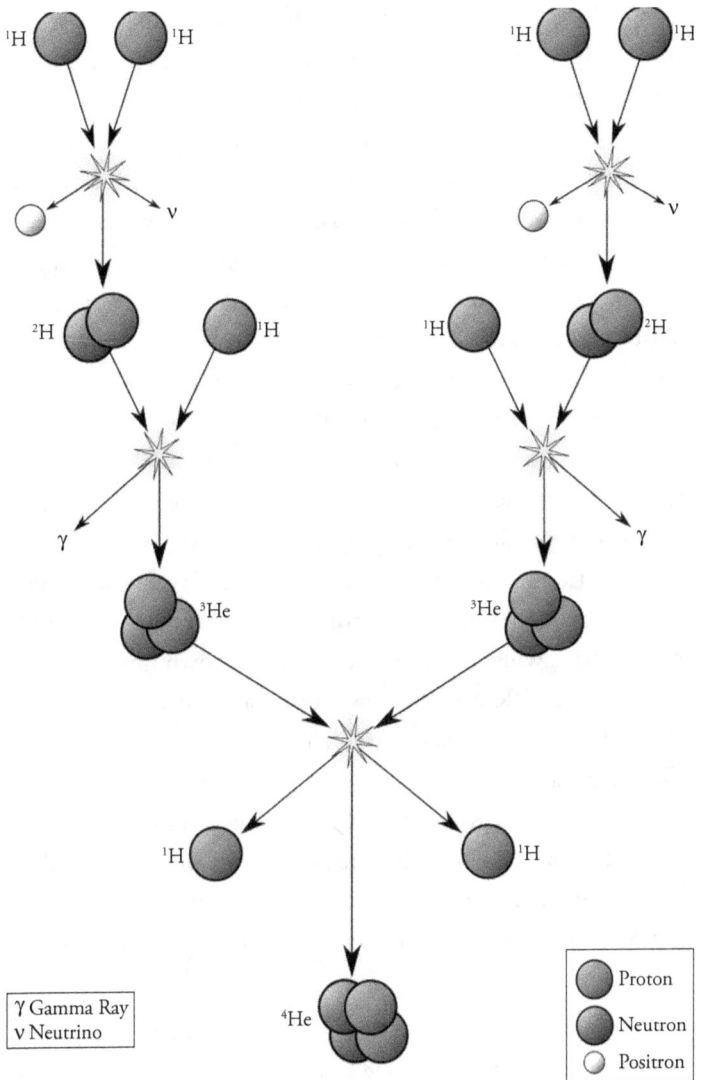

Fig. 2.2. *The proton-proton chain in the Sun. Note the production of neutrinos. Image courtesy of Wikimedia Commons.*

As the nickname ghost particle indicates, neutrinos are notoriously hard to detect as they interact very rarely with other particles, yet they pass through matter without much interaction. However, in 1956, Clyde Cowan and Frederick Reines discovered neutrinos and were awarded the Nobel Prize for their discovery in 1995. They used a nuclear reactor to produce neutrinos and detected them using photomultiplier tubes. Their experiments to detect the elusive neutrinos were originally conducted in Hanford, Washington, but they later moved the experiments to the Savannah River Plant near Augusta, Georgia where they had better shielding against cosmic rays. This shielded location was 11 m from the reactor and 12 m underground. Their discovery heralded a new era in physics, neutrino physics, that would eventually open a new field in exploring the universe. The unique ability of neutrinos to travel astronomical distances between their source origin and Earth made them ideal particles with which to learn about the unknown universe.

Cowan and Reines demonstrated that neutrino production occurred in nuclear reactors, which led to their first discovery: When particles transform into different particles, neutrinos are created in the process. Scientists identify neutrinos by detecting the fundamental particles they are associated with. There are three different types of neutrinos (electron, muon, and tau), each type relating to a charged particle as shown in the following Table 2.1.

Table 2.1. *There are three neutrino flavors (the electron neutrino, muon neutrino, and tau neutrino), which are related to the three charged particles (the electron, muon and tau). There could be additional flavors of neutrinos, but researchers are not sure about their existence.*

Neutrino	ν_e	ν_μ	ν_τ
Charged Partner	electron (e)	muon (μ)	tau (τ)

It is ironic that we don't know much about neutrinos, yet they are the second most abundant particle in the universe; they are everywhere. As mentioned earlier, their mystery is compounded by the fact that they appear in three types, or flavors, and shift among these flavors—electron, muon, and tau. As they travel, an electron neutrino can change into a muon neutrino or tau neutrino. Researchers believe that the study of neutrinos holds the key to knowing more about mysterious astrophysics sources, such as gamma-ray bursts, supernova explosions, and cataclysmic phenomena involving black holes and neutron stars. Recently, scientists have added another layer to the mystery surrounding these particles as they debate the idea of a fourth type of neutrino. Sterile neutrinos could only interact gravitationally, and that might shed light on the questions regarding neutrino mass and the role they play in understanding the existence of dark matter in the universe.

Given their ability to travel across the universe carrying momentum and energy, these ghost particles are unique and can help researchers explore the unknown. However, these particles are much stranger than scientists initially thought. To understand that, we need to know a little about antiparticles. The idea of the antiparticle came about in 1928 when British physicist Paul Dirac developed what became known as the Dirac equation. In 1932, physicist Carl Anderson discovered the anti-matter partner of the electron that Dirac had theorized earlier. This particle is called the positron—a particle like an electron but with a positive charge. Since these studies and the detection of antiparticles, it has been assumed that every particle has an antiparticle, although some have yet to be discovered. Here is the strange twist with neutrinos. In 1937, Italian physicist Ettore Majorana posited another theory: Neutrinos and antineutrinos are actually the same thing. He theorized that neutrinos, if they happened to have mass after all, could turn into antineutrinos and then back into neutrinos again. This idea has not been confirmed by experiments so far.

Now, let us go back to our initial question to see the connection between neutrinos and the existence of our material universe.

Why does anything exist? The largest particle physics experiment since the Large Hadron Collider may answer the question

Why does anything exist? We all ponder this question. Despite all the attempts to answer this question, our comprehension is no better than that of the ancient thinkers. Now an experiment, in fact the largest since the Large Hadron Collider, is poised to undertake this challenging question.

Scientists believe an asymmetry that occurred in the early universe is the root cause of everything, including our own existence. According to our current understanding, the Big Bang should have created matter and antimatter in equal amounts. When they interact, both matter and antimatter completely annihilate each other in a flash of radiation. However, in the early universe, matter somehow dominated the antimatter, resulting in a universe with galaxies, stars, planets, and even life.

The victory of matter over antimatter is an unsolved puzzle in modern physics. Many physicists suspect that the answer lies with neutrinos, the most abundant matter particles in the universe that pass through everything (including trillions of them through our body each second). They believe that heavy neutrinos and their antimatter counterparts, antineutrinos, populated the young universe but subsequently decayed. However, if this decay happened asymmetrically in the early universe, that would have created fewer antiparticles than particles. In other words, the difference in the decay rate of neutrinos and antineutrinos could hold the key to understanding why everything exists today rather than having been annihilated long ago.

On July 21, 2017, a groundbreaking ceremony was held at the Sanford Underground Research Facility in Lead, South Dakota to jumpstart the building of the Deep Underground Neutrino Experiment (DUNE), an international experimental facility to be completed over the next ten years which will be operated by 1,000 scientists and engineers from thirty countries (Figs. 2.3 and 2.4). This experiment will enable scientists to look for differences in the behavior of neutrinos and antineutrinos, which could provide essential clues as to why we live in a matter-dominated universe that supports life as we know it.

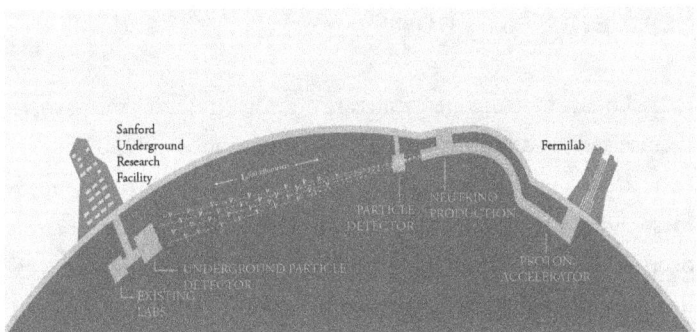

Fig. 2.3. *The international Long-Baseline Neutrino Facility/Deep Underground Neutrino Experiment, hosted by the U.S. Department of Energy's Fermilab, is an international flagship science project to unlock the mysteries of neutrinos. Image courtesy of Fermilab.*

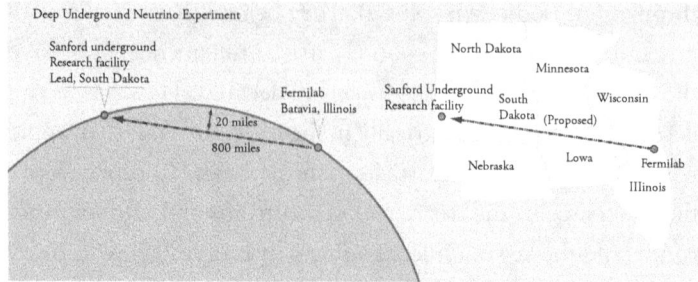

Fig. 2.4. *The Deep Underground Neutrino Experiment will send neutrinos 800 miles through the Earth. They will travel from Fermilab in Illinois to detectors one mile underground at the Sanford Underground Research Facility in South Dakota. Image courtesy of Fermilab.*

If the Large Hadron Collider was able to discover the God particle (Higgs boson), responsible for mass in the universe, the DUNE experiment has the potential to uncover the behavior of ghost particles (neutrinos) that would explain why anything exists.

Searching for ghosts

Experiments to detect neutrinos are often as strange as the particles themselves. In addition to the above-mentioned DUNE facility, there are several detectors around the world that study these ghostly messengers. One such facility is the IceCube Neutrino Observatory deep within the South Pole ice (Fig. 2.5). Approximately three hundred physicists from forty-nine institutions in twelve countries make up the IceCube Collaboration. Encompassing a cubic kilometer of ice, IceCube searches for nearly massless neutrinos that might come from distant corners of our galaxy or beyond.

Fig. 2.5. *The IceCube Laboratory at the Amundsen-Scott South Pole Station in Antarctica hosts the computers that collect raw data from the sensors buried in the ice below. Image courtesy of https://icecube.wisc.edu/about.*

This experimental facility, often called the Antarctic neutrino observatory, was designed as a multipurpose experiment. IceCube collaborators address several big questions in physics, like the nature of dark matter and the properties of the neutrino itself. IceCube also observes cosmic rays that interact with the Earth's atmosphere, which are not yet fully understood. In a recent report (July 2018), the researchers from the IceCube Collaboration provided evidence of a source of high-energy cosmic neutrinos billions of light-years from Earth.

For most of our existence, we humans have looked to the heavens and invoked the gods to help us make sense of the question, "Why do we exist?" Philosophers and scientists alike have pondered this question and come up with varied explanations. Despite all these attempts, it remains unanswered. The study of neutrinos might shed light on this eternal question, but will also add further questions like the one below.

Which came first—the laws of physics or the universe?

That sounds like the chicken-and-egg question. The prediction of neutrinos, their eventual discovery, and the ongoing research in detecting and analyzing other features of these ghostly particles all originated from a simple equation that was put forth to avoid the violation of the law of conservation of momentum, specifically the law of conservation of angular momentum. Naturally, we could pose this question: Which came first—the laws of physics or the universe?

Some theorists have beliefs similar to Plato's theory of ideas or forms, which postulated that ideas existed, and nature fell into those forms. Accordingly, mathematical formats existed even before the universe began, and we discovered them later. However, experimentalists disagree as they observe the behavior of the existing universe and try to fit those observations with the theoretical models.

We cannot answer this question without referring to time because, semantically, *first* is linked to time. In our current understanding, space and time originated with the Big Bang. So, the thinking of the theoretical physicists—that the laws existed before the observable universe—cannot hold since we are presuming that time is there and so the universe. An experimentalist could argue that our knowledge is limited by our experiments and observations, and again that needs the universe to be present. So, the debate over whether the laws of nature *describe* the universe or *govern* the universe will rage on. Such debates reveal the deep desire of the human mind to comprehend the unknown. In the meantime, you don't have many choices with the laws of nature. You can't disobey them.

While the above question may remain unanswered for now, physicists in general agree that the laws of physics are universal. One of the quantities that can verify the universal nature of the laws of physics is the fine structure constant α. This magic number is very closely equal to 1/137 and measures the strength of the electromagnetic force that governs how electrically charged elementary particles, like electrons and light (photons), interact. It can be expressed as a combination of three constants: the charge on an electron, e, the speed of light, c, and Planck's constant, h. It is believed that this fundamental constant does not vary wherever our observation happens in the universe.

Lately, researchers (King et al., 2012) have presented evidence for the variation of this constant based on measurements taken with the Very Large Telescope (VLT) in Chile. They reported that the value increases with increasing cosmological distance from Earth, although very slightly. This is an indication that the laws of physics may vary over cosmic time or distances (Fig. 2.6).

It is generally assumed that the laws of physics, which are often referred to as the laws of nature, must be consistently accurate beyond space and time constraints. But what if they evolve over time like any other entity? Finally, there could be laws of nature that we are unaware of, and they might be operating in hidden dimensions. Although it would be ideal to consider that the laws of physics and physical constants are the same everywhere and all the time, no restrictions can be imposed on them to require such a uniformity. I am sure scientists have to face up to that possibility as we study more about these laws.

There could be cases where the seemingly rigid pieces that make up the science change as we have just seen with the fine structure constant. This calls to mind what the French mathematician, Henri Poincaré, noted in *Science and Hypothesis*, "Science

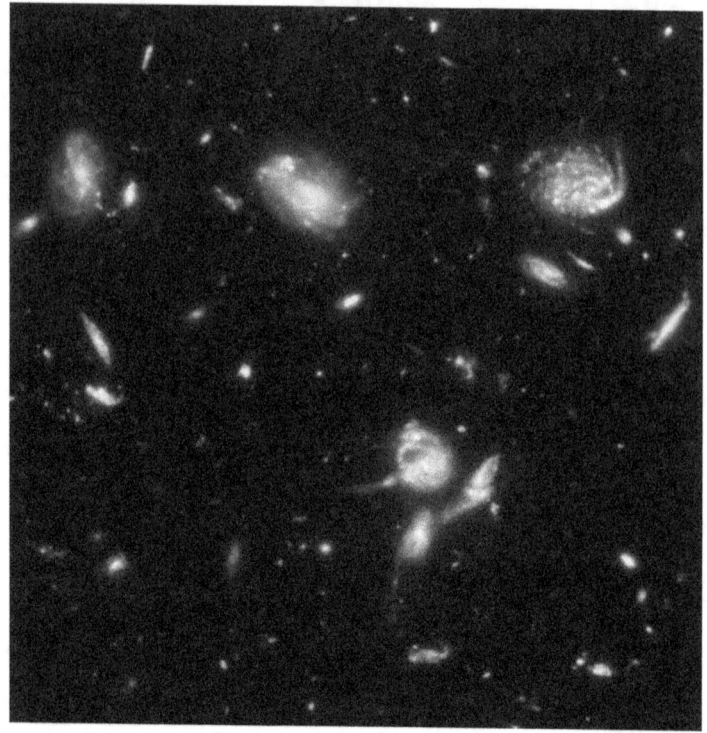

Fig. 2.6. *Some researchers report they have evidence of changes in one of the fundamental constants of nature. Perhaps we are living in a section of the universe that is "just right" for our existence. Image courtesy of NASA.*

is built up with facts, as a house is with stones. But a collection of facts is no more a science than a heap of stones is a house."

We believe what we see, but science has long taught us that our eyes can deceive. What we see today might be the drama of the cosmos that unfolded long ago. The seemingly different past, present, and future lose their meaning in the vastness of the universe. Contrary to popular belief, physicists rejoice when the laws of nature are violated. That provides them an opportunity to delve even more deeply into the secrets of nature. The imperishability of the equations can be questioned, at least in some cases, but their ability to connect various phenomena in the universe remains eternal.

Bibliography

Aartsen, M. G., et al. (2018, July 13). Multimessenger observations of a flaring blazar coincident with high-energy neutrino IceCube-170922A. *Science, 361*(6398). doi: https://doi.org/10.1126/science.aat1378

Ball, P. (2012, March 29). Can the laws of physics change? *BBC Future.* Retrieved from http://www.bbc.com/future/story/20120329-can-the-laws-of-physics-change

Baggott, J. E. (2013). *The quantum story: A history in 40 moments.* Oxford: Oxford University Press.

Brewster, S. (2016, January 20). Is the neutrino its own antiparticle? *Symmetry.* Retrieved from https://www.symmetrymagazine.org/article/is-the-neutrino-its-own-antiparticle

Castelvecchi, D. (2019, February 27). Gigantic Japanese detector prepares to catch neutrinos from supernovae. *Nature.* Retrieved from https://www.nature.com/articles/d41586-019-00598-9

Fermilab. (2017, June 15). The Science of the Deep Underground Neutrino Experiment (DUNE). Retrieved from https://www.youtube.com/watch?v=nv13DswIKr8

Fermilab. Neutrinos. (2017, November 4). Saturday morning physics. Retrieved from https://indico.fnal.gov/event/14559/material/slides/6.pdf

Fermilab News. (2017, July 21). Construction begins on international mega-science experiment to understand neutrinos. Retrieved from http://news.fnal.gov/2017/07/construction-begins-international-mega-science-experiment-understand-neutrinos/

Gardner, M. (1990). *The new ambidextrous universe: Symmetry and asymmetry from mirror reflections to superstrings.* New York, NY: W.H. Freeman.

Ghosh, P. (2014, February 14). UK backs huge US neutrino plan. *BBC News*. Retrieved from https://www.bbc.com/news/science-environment-26017957

Giunti, C., & Kim, C. W. (2007). *Fundamentals of neutrino physics and astrophysics*. Oxford: Oxford University Press.

Heisenberg, W. (1971). *Physics and beyond: Encounters and conversations*. New York: Harper & Row.

Huber, J., & Jepsen, K. (2015, March 25). The dawn of DUNE. *Symmetry*. Retrieved from https://www.symmetrymagazine.org/article/march-2015/the-dawn-of-dune

IceCube Collaboration (n.d.). Retrieved from https://icecube.wisc.edu

Introduction to the constants for nonexperts (n.d.). The NIST Reference on Constants, Units and Uncertainty. Retrieved from https://physics.nist.gov/cuu/Constants/alpha.html

Jayawardhana, R. (2015). *The neutrino hunters: The chase for the ghost particle and the secrets of the universe*. London: Oneworld.

King, J. A., Webb, J. K., Murphy, M. T., Flambaum, V. V., Carswell, R. F., Bainbridge, M. B., …. Koch, F. E. (2012). Spatial variation in the fine-structure constant—new results from VLT/UVES. *Monthly Notices of the Royal Astronomical Society, 422*(4), 3370–3414. doi:10.1111/j.1365-2966.2012.20852.x

Lederman, L. M., & Hill, C. T. (2013). *Beyond the God particle*. Amherst, NY: Prometheus Books.

Lederman, L. M., & Teresi, D. (1993). *The God particle: If the universe is the answer, what is the question?* Boston: Houghton Mifflin.

Neutrinos from Beta decay. (n.d.). Retrieved from http://neutrinos.fnal.gov/sources/beta-decay/

Pauli, W., Jung, C. G., Meier, C. A., Enz, C. P., Fierz, M., & Roscoe, D. (2014). *Atom and archetype: The Pauli/Jung letters, 1932–1958*. Princeton: Princeton University Press.

Poincaré, H. (1952). *Science and hypothesis*. New York: Dover.

Swinburne University of Technology. (2010, September 9). Laws of physics vary throughout the universe, new study suggests. *ScienceDaily*. Retrieved February 3, 2019 from www.sciencedaily.com/releases/2010/09/100909004112.htm

CHAPTER 3

The equation that launched rockets and brought down apples

Because there is a law such as gravity, the universe can and
will create itself from nothing.
—*Stephen Hawking and Leonard Mlodinow,*
The Grand Design (2010)

Abstract

While the story of the apple falling on Newton's head is probably a
myth, he figured out why it fell. Newton's law of gravitation was
first published in 1686, but even today scientists and engineers use
it to design rockets and aircraft carriers. Although Einstein's general
theory of relativity revealed a new way of looking at gravity, the
equation attributed to Newton still bears relevancy to the astronom-
ical sciences today. The path of a spacecraft and the orbit of every

satellite in space can still be calculated using this simple formula, and it works magnificently.

Among all the four fundamental forces (gravity, electromagnetism, and strong and weak nuclear forces), gravity is the one most obvious to us and is considered to be well understood. Yet, many aspects of gravitational force remain to be explored, and it is arrogance on our part to make the claim that we comprehend it completely. If we ever want to discuss the force of gravity, I think we should begin with Newton's law of universal gravitation:

$$F = G \, mM \, / \, r^2$$

G – gravitational constant
m – mass of the first object
M – mass of the second object
r – the distance between objects

In his own words, Newton defined the force of gravity in the following way:

> Every particle of matter in the universe attracts every other particle with a force that is directly proportional to the product of the masses of the particles and inversely proportional to the square of the distance between them.

The brilliance with which Newton expressed this relationship, as shown in the above equation, is a testament to his genius.

However, the popular notion that Isaac Newton discovered gravity is an ambiguous statement. Of course, his equation describes the force of gravity in simple terms and is considered as one of the most elegant equations in physics, but people thought about gravity well before Newton. However, his equation is remarkable in the sense that it conveys a powerful message that every material object attracts everything else in the universe, and this force of attraction depends only on the mass and distance between the objects involved. This universal nature of gravity, as shown by Newton's equation, links earthly bodies to heavenly objects as it teaches us that the force that keeps the Moon orbiting around the Earth is the same force that makes an apple fall to the ground. This is a remarkable way of connecting the seemingly different phenomena with the same underlying principle.

The force of gravity is unique in the sense that it differs from other fundamental forces mentioned earlier, and it is the force with which we are so familiar since it is tied to our own existence. It keeps the planets around the star and the moons around the planets, thus providing stability and order for this universe. In the absence of this gravitational force, inertia will keep a planet going in a straight line instead of a circular orbit. Gravity also has a significant role in dictating how everything stays in place in this universe, including our own existence on this planet, and when we venture into space, noticeable changes happen to the human body in the absence of normal gravity.

We know that the strength of gravity is six times smaller on the Moon than on Earth. A lunar astronaut who weighs roughly 150 pounds on Earth would weigh only 25 pounds on the Moon. It becomes difficult for individuals on the Moon to firmly keep their feet planted on the surface, but it becomes easier to leap a couple of feet off of the surface. While in space, the human body changes in a plethora of ways, and that can affect the long-term health of the person. Due to the lack of gravity pulling down on the body, the bones and muscles work less; without physical activity, the result is a prolonged level of atrophy. Fluids also tend to float around, and as a result, the fluids tend to float up to the head without gravity to help hold them down. Cosmic radiation also affects human health at a microscopic level by damaging DNA and RNA, which can cause cancer or other genetic diseases years down the line. The body is impacted massively, even in a matter of months, in Low Earth Orbit (LEO). But none of these will stop the human species from venturing into space. After all, it has been fifty years since humans landed on the Moon for the first time.

Scientists and engineers depend on Newton's gravitational equation to launch the rockets that take humans into space. Human fascination with space exploration took a giant leap when Apollo 11 landed men on the Moon in July 1969 (Fig. 3.1). Since that most celebrated event in human history, humankind has not made any huge strides to further extend this revolutionary achievement. Our interactions with Earth and space reveal that we are still a planet-dependent species that pollutes the

Fig. 3.1. *On July 16, 1969, the huge, 363-feet tall Saturn V rocket launches on the Apollo 11 mission from Kennedy Space Center. Newtonian law predicts the path of a rocket based on the calculations involving the universal gravitation equation. Image courtesy of NASA.*

environment to the brink of destruction; we are not a spacefaring civilization. In many ways, the last five decades since the Moon landing have been lost decades, especially when it comes to space explorations. These years could have been used as a stepladder to become an elevated species that draws its resources and energy from a star, not from a mere planet—an advanced civilization. Recently, many blueprints have been floated to take humans to other planets, especially to Mars. The strength of the gravitational equation will be tested further when humans venture out to other celestial bodies beyond the Moon or Mars.

Newton's theory of gravity, published in 1687, provided a universal approach in calculating the force between different objects regardless of their size and shape. Even today, it remains a force to be reckoned with although Einstein's theory of general relativity provides a very different picture of gravity that Newton did not envision. Ideally, Newton's law would work for two bodies but, when there are more interactions, it is not easy to figure out the forces and subsequent motion of the objects involved, and modern computer simulations become necessary for those tasks. In addition to that, when we need to deal with the deep questions of modern-day astronomy, such as the expanding universe or black holes, we need to depend on the theory of relativity rather than Newton's law.

Since its publication in 1915, general relativity has been acknowledged as the theory that best explains gravity, and it has superseded Newtonian mechanics that reigned over the world of physics for almost 200 years. In the general theory of relativity, gravitation is not viewed as a simple force, but rather a property of the space-time curvature. Space and time, in Einstein's universe, are no longer flat (as implicitly assumed by Newton) but can be pushed and pulled, stretched and warped by matter. Gravity feels strongest where space-time is most curved, and it vanishes where space-time is flat. This is the core of Einstein's theory of general relativity, which is often summed up in the words of legendary physicist John Wheeler as follows: "matter tells space-time how to curve, and curved space-time tells matter how to move." This theory transformed astronomy in addition to physics. Astronomers now have a new way of explaining many phenomena that baffled them in the past.

Newton's view of the universe was a grand clockwork in which material objects played a role set by the creator. He believed gravity worked instantaneously across the universe, which is refuted by the Einsteinian view of the universe (Fig. 3.2). The general theory of relativity proclaimed that the gravitational interactions between massive objects generate waves aptly called

Fig. 3.2. *Einstein's theory of general relativity predicted that the space-time around Earth would not only be warped but also twisted by the planet's rotation. Image courtesy of NASA.*

gravitational waves and they travel at the speed of light across the universe. In other words, gravity cannot act instantaneously as Newton thought. We all are familiar with the force of gravity, but now it looks more complicated than the simple picture painted by Newton.

So, where do the gravitational waves fit in the picture? These waves should stretch and compress the space-time fabric when they pass through the universe as predicted by Einstein's theory. However, even Einstein doubted we would ever be able to detect such a signal. Many attempts have been made to identify gravity waves since the 1970s, but they were detected only in 2017, almost one hundred years after they were predicted to exist by Einstein's theory.

One hundred years of gravitational waves

So, here is the basic difference between the Newtonian theory of gravity and Einstein's general theory of gravity. Although Newton's equation still works as a fundamental equation that enables us to calculate the force of gravity, his assumption that gravity acts instantaneously between two masses, even if the two objects are separated by light-years, does not make much sense. The special theory of relativity tells us that nothing can travel faster than the speed of light. Thus, two stars cannot interact instantaneously since the gravitational waves must first reach the other object for the interaction to occur. This is the essence of the general theory of relativity. As mentioned above, researchers have been looking for evidence of gravitational waves since their prediction, but the detection remained elusive for a long time.

The experimental setup that was able to detect gravitational waves is called Laser Interferometer Gravitational-wave

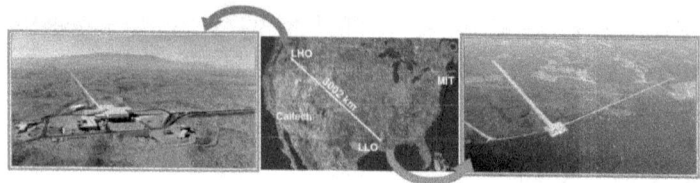

Fig. 3.3. *The general locations of the LIGO Hanford and LIGO Livingston interferometers. Image courtesy of Caltech/MIT/LIGO Lab.*

Observatory (LIGO), the world's largest gravitational wave observatory, and it is an engineering marvel (Fig. 3.3). Comprising two enormous laser interferometers located thousands of kilometers apart, LIGO exploits the physical properties of light and of space itself to detect and understand the origins of gravitational waves (Fig. 3.4). LIGO is jointly operated by Caltech and MIT and is

Fig. 3.4. *Gravitational Waves of a compact binary system. Image courtesy of MoocSummers, Wikimedia Commons.*

supported by the U.S. National Science Foundation. On September 14, 2015, at 5:51 a.m. ET, LIGO detected gravitational waves for the first time. This discovery confirmed one of the major predictions of Einstein's general theory of relativity, and it has been assumed that this discovery will definitely pave the way for a new era of observational astrophysics.

As mentioned earlier, the general theory of relativity, published one hundred years ago, describes space-time as a single dynamic entity and predicts bizarre phenomena, like black holes and gravitational waves. In other words, Einstein discarded the Newtonian notion of the independent nature of space and time and argued that space was not a mere canvas upon which all events unfolded by the mysterious force of gravity. The theory of relativity also predicts that a strong gravitational field, similar to that of the Sun, "warps" space-time and that even light is deflected by gravity. In 1919, the British astrophysicist Arthur Eddington measured the bending of starlight around the Sun during a solar eclipse as predicted by the general theory of relativity. It was a clear vindication of the validity of this theory. There have been other verifications of the general theory of relativity since the days of Eddington.

In fact, the seemingly abstract concept of the warping of space-time produces noticeable effects, such as time dilation that has been measured precisely. The general theory of relativity predicts that the warping of space-time will be larger in stronger gravitational fields. Apparently, clocks on the ground (closer to Earth) will run more slowly than the clocks, say, aboard the GPS satellites orbiting the Earth. Einstein's equations calculated this time difference to be about 38 microseconds per day, and these relativistic effects were taken into account by adding the required mathematical corrections into the system when the GPS satellites were deployed. The general theory of relativity, linked to the special theory of relativity, explains even such microscopic anomalies about space and time so beautifully and together they form the foundation of modern physics.

Physicists often describe gravitational waves as "ripples on space-time" caused by moving masses similar to the electromagnetic waves caused by moving charges. Just as a pebble dropped in a pond can generate waves propagating in all directions across the water, cataclysmic events, such as the collisions of black holes, will send out energy propagating through the fabric of space-time in the form of gravitational waves. LIGO scientists estimate that this particular event, detected on Earth in September 2015, took place about 1.3 billion years ago. They confirm that two massive black holes generated these waves as they spiraled against each other until they merged together, sending out ripples that washed across our planet at the speed of light. Indeed, these waves are the voices of stellar ghosts (black holes) as they ripped each other apart in the far corners of the universe, and that message was sent to us over the cosmic network.

Astronomers rely on different forms of electromagnetic waves to observe the universe. For example, radio telescopes are used in astronomy to explore the universe by detecting radio signals emitted by different objects. However, we learned to communicate with them—think of mobile communication that is entirely dependent on radio waves. The detection of gravitational waves marked a triumphant moment in scientific history and will provide a completely new way of looking at the universe. What will these bring to physics in coming years? A new wave from the cosmic ocean, or perhaps an ocean itself?

What goes up must come down—really?

People often invoke Newton by saying what goes up must come down. Although not very accurate, they are safe in making that argument as there is no mention of time.

When it comes to satellites, engineers can make satellites come down or go further up from their actual orbit, as needed. In the early days of space exploration, scientists and engineers did not have to worry about the fate of the satellites they had put into space once they became inoperative. Normally, smaller satellites that circled the Earth in low orbits burned up as they entered the Earth's atmosphere after years of space duty, and no one noticed them. However, that is not the case with massive satellites even if they are in lower orbits.

What do the spacecraft operators do with larger objects like space stations in low orbit? These objects may not entirely burn up before reaching the ground due to their size. The only solution for now is to choose a final falling destination in a remote area where debris won't affect normal life. This location has a nickname—the Spacecraft Cemetery! It's in the Pacific Ocean and is pretty much the farthest place from any human civilization you can find (Fig. 3.5).

For the satellites high above the Earth, this technique won't work as it requires a huge amount of fuel, so instead of applying brakes to its orbital speed, the satellite is pushed into a higher orbit, known as the "graveyard orbit." Also referred to as "junk orbit" or "disposal orbit," it lies 36,050 km above Earth. Pushing dead satellites into the junk orbit requires much less effort than de-orbiting them. Thus, the dead satellites are given a second life as they continue to orbit

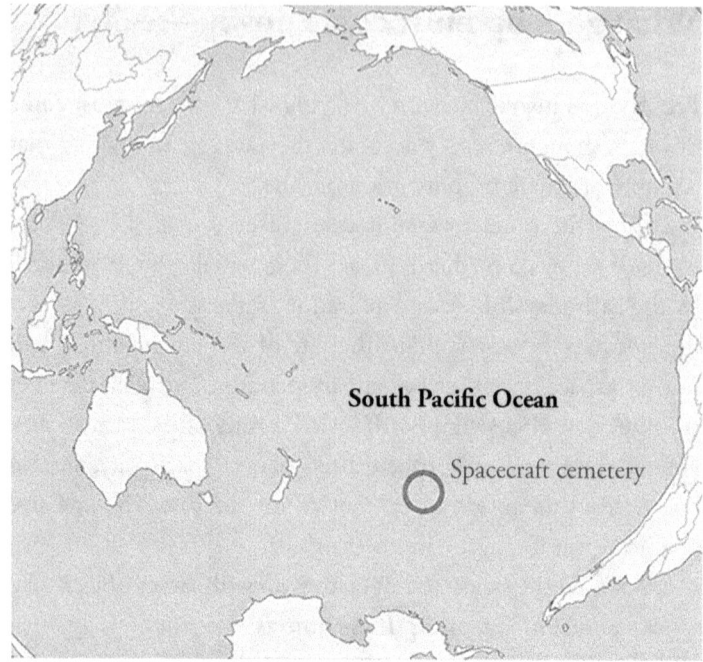

Fig. 3.5. *Spacecraft Cemetery in the South Pacific Ocean, far from where anyone lives. Image courtesy of NASA.*

the Earth. Space agencies have placed hundreds of satellites in their retirement home, and it works for the time being. However, in the future, a high-orbit collision is a possibility as more and more satellites are pushed up, and they could pose a danger to active satellites. The takeaway from this discussion is that the saying, *what goes up must come down*, is not a well-thought-out statement. We can definitely conclude that certain popular notions are accepted without being subjected to the rigorous standards of science. For example, even in the twenty-first century, there are people who seriously believe astrology can predict our future.

Gravity and the role of the creator

Both philosophers and physicists have debated the role of a creator in the material universe for a long time. In his 1988 book, *A Brief History of Time*, it appeared that Stephen Hawking (Fig. 3.6) had indicated the role of God in the creation of the universe. However, in his later book, *The Grand Design* (2010), co-written with American physicist Leonard Mlodinow, he claimed that the laws of physics can explain the creation of a material universe, and a creator is not necessary. He wrote, "Because there is a law such as gravity, the universe can and will create itself from nothing."

Theologians would try to negate that argument by saying the existence of a creator is, by definition, outside the domain of science. So, what about the laws of physics? Do the laws of physics predate the Big Bang that, scientifically, we all agree marked the beginning of our universe? These questions are debated in physics and philosophy with no clear outcome and will continue to remain active in the minds of thinkers for years to come.

Isaac Newton held the belief that God must have designed the universe, as it could not have been created from nothing. Perhaps this explains why he never bothered to explore how the universe worked although his equations make perfect sense even today. Albert Einstein once wondered "whether God had any choice in the creation of the world." To Einstein, God was a code word for the mystery and grandeur of the universe, the wellspring of awe, a reminder that there was something at the core of existence that all his equations could only scratch the surface of, "something we cannot penetrate" (Overbye, 1999). It is well recognized that Einstein used to mention God more frequently than you would expect for a scientist, mostly in relation to the

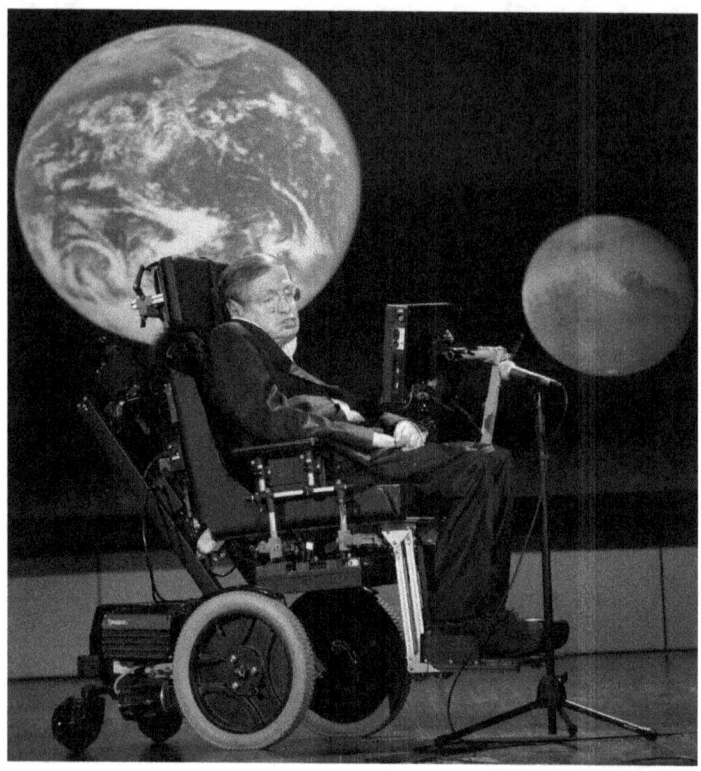

Fig. 3.6. *Stephen Hawking, theoretical physicist, cosmologist, and a former professor of mathematics at the University of Cambridge, delivered a speech entitled "Why we should go into space" during a lecture that was part of a series honoring NASA's 50th Anniversary, Monday, April 21, 2008, at George Washington University's Morton Auditorium in Washington. Hawking died on March 14, 2018. Image courtesy of NASA/Paul E. Alers.*

design of the universe. But that had nothing much to do with the picture of God painted by religious texts which were, in Einstein's own words, "an incarnation of the most childish superstitions."

It has been estimated that our own species, Homo sapiens, distinguished by bigger brains, more tool-making skills and the ability to reach far beyond, arose about 200,000 years ago and managed to survive and thrive despite climate change at the time. Later on, the first humans began straying outside of the continent in which our species was born. Our scientific pursuit began only recently (a few hundred years ago) on a planet that is itself considered to be 4.5 billion years old. We are not at the end of the cycle yet, and there is no guarantee that we are or ever will be able to unlock the cosmic mysteries through science, but that is the best bet we have at the moment. And what makes us unique is our ability to explore further and experience the joy of the scientific process regardless of the outcomes.

Our picture of a creator or God is shaped by the stories in the texts written during only a short period of human evolution, and there is no reason to think that all comprehensive knowledge resides in those books. Even modern interpretations of those texts only invite us back to a stagnant pool where medieval values are in charge. These texts are, at best, honorable legends, and as Hawking said, "It is not necessary to invoke God to light the blue touch paper and set the universe going." If we have to label anything as God, it should be the laws of nature.

Isaac Newton died on March 20, 1727 and was buried in Westminster Abbey. Not surprisingly, in 2018 when Stephen Hawking died, his ashes were buried near the graves of Charles Darwin and Isaac Newton and later covered with a gravestone. Our picture of gravity has improved since Newton came up with his famous equation, and it could happen again in the future. Humans will depart this Earth, and our old ideas will be superseded as science marches forward, yet the equation that provided us with a way to quantify gravitational force will remain eternal.

Bibliography

Einstein, A. (1922). The general theory of relativity. In *The meaning of relativity*. Springer: Dordrecht.

Einstein, A., & Lawson, R. W. (1921). *Relativity: The special and general theory*. New York: Holt.

Gravitational Waves. (n.d.). Retrieved from https://www.ligo.caltech.edu/page/gravitational-waves

Hawking, S. (1998). *A brief history of time*. New York: Bantam Books.

Hawking, S., & Mlodinow, L. (2010). *The grand design*. New York: Bantam Books.

Hawking, S. W., & Ellis, G. F. R. (1975). The large scale structure of space-time (Cambridge Monographs on Mathematical Physics). Cambridge University Press.

LIGO. Retrieved from https://www.ligo.caltech.edu/page/about

Newton, I., & Leedham-Green, C. R. (2018). *The mathematical principles of natural philosophy: An annotated translation of the Principia*. Cambridge: Cambridge University Press.

Overbye, D. (1999). Did God have a Choice? *The New York Times Magazine*. Retrieved from http://movies2.nytimes.com/library/magazine/millennium/m1/overbye.html

Relativity, Special Theory of. Encyclopedia of Science and Religion. Retrieved January 28, 2019 from Encyclopedia.com: https://www.encyclopedia.com/education/encyclopedias-almanacs-transcripts-and-maps/relativity-special-theory

Westminster Abbey. (n.d.). Sir Isaac Newton. Retrieved from https://www.westminster-abbey.org/abbey-commemorations/commemorations/sir-isaac-newton

Westminster Abbey. (n.d.). Stephen Hawking. Retrieved from https://www.westminster-abbey.org/abbey-commemorations/commemorations/stephen-hawking

CHAPTER 4

The equation that tells galaxies are flying away from us

The history of astronomy is a history of receding horizons.
—*Edwin Hubble, The Realm of the Nebulae (1936)*

Abstract

It is surprising to believe that a simple algebraic equation can have such a profound impact in modern cosmology. In 1929, based on his observations of the galaxies, Edwin Hubble announced that almost all galaxies appeared to be moving away from us. This phenomenon was observed as a redshift of the galaxies' spectrum similar to the Doppler Effect in sound where a receding source of waves will create an apparent change in frequency to an observer. So, in billions of years we will be left on our own island galaxy with all our neighbors having departed to the far corners of the cosmos.

The observational evidence gathered by Edwin Hubble in the 1920s culminated in one of the greatest discoveries of the last century. Hubble's discovery transformed the picture of our cosmos in profound ways. It changed not just the way astronomers viewed our place in the universe, but it also changed the human perception of the universe. As is well known, Hubble's work led the scientific community to abandon the idea that our universe had just one galaxy named the Milky Way. His greatest discovery came in 1929 when he determined that the farther a galaxy is from Earth, the faster it appears to move away—a new paradigm shift in our understanding. In other words, we live in an "expanding" universe which implies that our universe began at a single moment in time and, of course, has been expanding ever since. The currently accepted scientific theory of the Big Bang as the origin of the universe has its roots in Hubble's findings.

In the 1920s, Hubble had access to the largest telescope of the day, situated on Mount Wilson, California. Using this 100-inch telescope, he found that nearly all the galaxies in the universe were moving away from us. This motion, called their recession velocity, was greater the further away they were from us. In fact, he found a simple linear relationship between a galaxy's velocity (v) and its distance from the Earth (d), which later came to be known as Hubble's law:

$$v = H_o\, d$$

where

v = velocity of a galaxy, in km/s

H_o = Hubble Constant, measured in km/s/Mpc (megaparsecs[1])

d = distance of a galaxy, in Mpc

---

[1] A megaparsec is a measurement of distance equal to one million parsecs or 3.26 million light-years. Megaparsec is usually abbreviated as Mpc.

It should be mentioned here that, for nearby galaxies or stars within the Milky Way, this relationship may not hold. Originally, Hubble estimated the value of the expansion factor called the Hubble constant (H_o) to be about 500 km/sec/Mpc. Today, astronomers still debate the precise value of the Hubble constant, generally considered to be in the range of 45–90 km/sec/Mpc (more on that discussion in coming sections). With this single discovery, Hubble revolutionized the long-held body of human knowledge about the universe and its eventual fate. Edwin Hubble died on September 28, 1953, and in memory of his great work, NASA named the world's first space-based optical telescope after him (the Hubble Space Telescope, abbreviated HST).

While, in general, galaxies follow the smooth expansion indicated by the Hubble equation—the more distant ones move faster from us—other motions, however, cause slight deviations from the line predicted by Hubble's law. The diagram (Fig. 4.1) shows a typical plot of distance versus recessional velocity, with each point showing the relationship for an individual galaxy.

In chapter 3, we saw how Einstein's theory of general relativity deviated from the Newtonian picture of the universe, especially over large scales. Similarly, for long distances, Einstein's theory predicted departures from a strictly linear Hubble law, and the extent of the departure depended on the value of the total mass of the universe. It should be noted here that, over a decade before Hubble made his observational discovery, the theory of general relativity offered the theoretical framework for an expanding universe. However, Einstein removed such a scenario from his theory owing to the fact that the existing evidence contradicted this radical idea. Later on, Einstein himself remarked that this was "my biggest blunder." In fact, Einstein visited Hubble at Mount Wilson after Hubble reported his observations, another illustration of how significant this finding was.

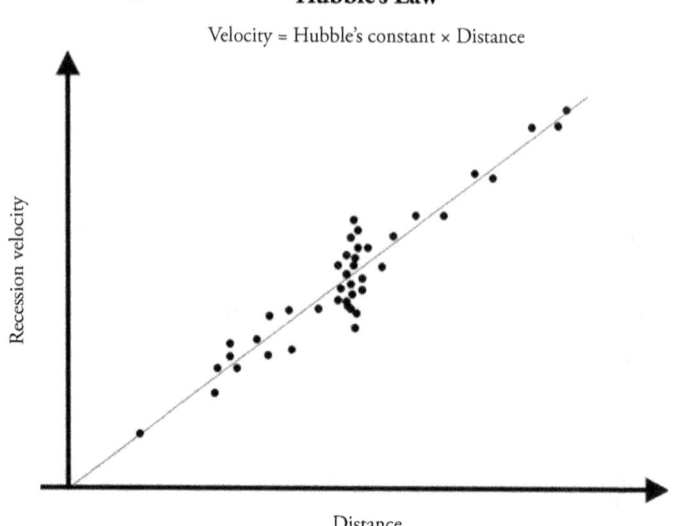

Hubble's Law

Velocity = Hubble's constant × Distance

Fig. 4.1. *Plot of recession velocity versus distance for galaxies, showing the basis of Hubble's law. The further away a galaxy is, the faster it recedes from us. Therefore, if we know how fast a galaxy is moving, we can determine its distance using this plot. Image courtesy of NASA.*

No doubt the profound impact of this linear relationship proves the majesty of Hubble's equation. No other simple equation has changed the view of our universe in such a radical fashion. To get a better sense of this equation, imagine the Hubble constant (H_0) is about 70 km/sec/Mpc (as mentioned above, the exact value of the Hubble constant is still debatable). This means that a galaxy 1 megaparsec away from us will be moving away from us at a speed of 70 km/sec. It is simple to extend this argument because the relationship between speed and distance is linear. Therefore, another galaxy that is 100 megaparsecs away from us will be cruising at 100 times faster than the galaxy that is 1 megaparsec away from us. The bottom line is that the

Hubble constant represents the speed at which the universe itself is expanding.

However, the above argument can lead us to an unpleasant situation in physics. If we continue to calculate the recessional velocity of galaxies farther and farther away, we will reach a tipping point where the velocity becomes even faster than the speed of light. How is it possible to have a speed greater than the speed of light? Eisenstein's special theory of relativity (in fact, the special theory of relativity was published before the general theory of relativity) forbids any such act, and it now looks like the expansion of the universe is violating one of the ironclad laws of physics.

The answer for this question lies in the difference between the expansion of the universe, which is essentially expanding the space-time fabric, and the motion of an object that was the focus of the special theory of relativity. However, it makes better sense if you understand the distinction between expansion and motion. When scientists say that the universe is expanding, they are talking about the rather abstract concept of space-time which includes the three physical dimensions of our existence, length, breadth, and depth, combined with the additional dimension of time (Fig. 4.2). We can imagine this space-time as a grid that connects every part of the universe to every other part. When we say an object is moving, we are referring to its change in position relative to the space-time grid. In other words, the speed of light is a constraint only for objects that exist within space-time, not for space-time itself. Thus, scientists resolve this apparent dispute by clarifying the difference between space-time expansion and the motion of an object. They cannot be treated the same way.

To understand this rather intricate problem of the expansion of the universe, we can depend on the above-mentioned

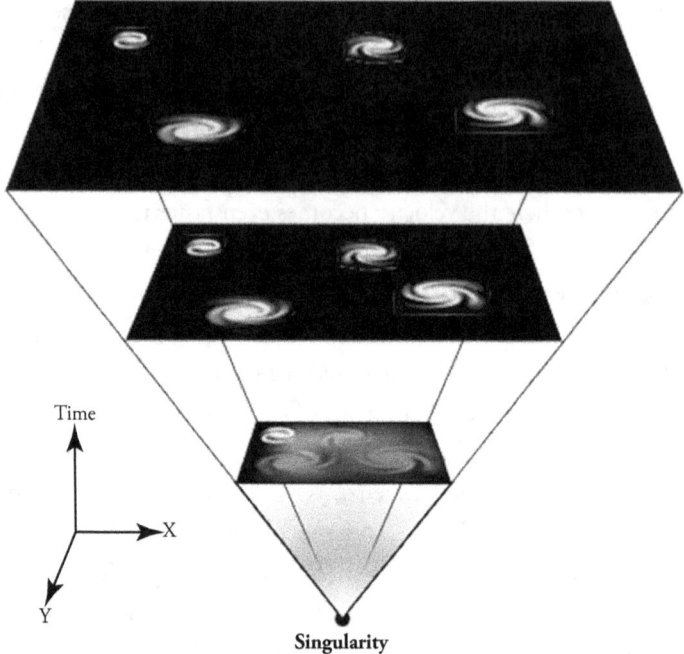

Fig. 4.2. *This illustration shows abstracted "slices" of space at different points in time as the universe expands. It is simplified as it shows only two of three spatial dimensions, to allow the time axis to be displayed conveniently. Image courtesy of Wikimedia Commons.*

Hubble equation. If you look carefully at the equation, the variable known as the Hubble constant (H_o) is the key here, and so let us focus on it for the time being.

How do astronomers actually measure the value of the Hubble constant?

In modern astronomy, the precise value of the Hubble constant (H_o) is of great interest as it determines the age of the universe, which is currently estimated to be around 14 billion years. Although the equation involving the Hubble constant is simple and linear, obtaining an accurate value for the Hubble constant is a very complicated process. From the equation, $v = H_o d$, it is clear that astronomers need two measurements. First, from spectroscopic observations, they should get the galaxy's redshift, indicating its radial velocity. The second measurement, the more difficult one to determine, is the galaxy's precise distance from the Earth. Astronomers use variable stars (Cepheid variables) and supernovae as indicators in the galaxies to determine the distance. Additionally, the value of Ho itself must be carefully derived from a sample of galaxies that are far enough away that motions due to local gravitational influences are negligibly small and can be ignored.

Two terms need special attention here to understand the calculation of the Hubble constant. First, let us talk about redshift. It is similar to what happens to sound waves when a source of sound moves relative to an observer. This effect is called the "Doppler effect" after Christian Andreas Doppler, an Austrian mathematician who discovered that the frequency of sound waves changes if the source of the sound and the observer are moving relative to each other. We can find several everyday examples of the Doppler effect—the changing pitch of police and ambulance sirens or train whistles and racing car engines as they pass by. In every case, there is an audible change in pitch as the source approaches and then passes an observer. Similarly, light behaves like a wave, so light from a luminous object undergoes a Doppler shift if the source is moving relative to an individual, away or towards. Ever since 1929 when Edwin Hubble discovered that the universe was expanding, we have known that most other galaxies are moving away from us. Light from these galaxies is shifted to longer (and this means redder) wavelengths; in other words, it is "redshifted."

Since light travels at such great speed relative to everyday phenomena (300,000 km/s, that is roughly a million times faster than sound), we do not experience this redshift with our own eyes. Usually, astronomers measure redshift of a distant galaxy or quasar[2] by comparing its spectrum with a reference laboratory spectrum. The spectrum consists of atomic emission and absorption lines that occur at well-known wavelengths for each chemical element. By measuring the apparent change of these lines in astronomical spectra, astronomers can determine the redshift of the receding sources. It must be clearly distinguished here that Doppler shifts arise from the relative motion of source and observer through space, whereas astronomical redshifts are due to the expansion of space itself.

The second term we need to understand is Cepheid variables or variable stars. These types of stars undergo a rhythmic pulsation as indicated by their regular pattern of changing brightness as a function of time. The period of pulsation is directly related to their intrinsic brightness, making observations of these stars one of the most powerful tools for determining distance known to modern-day astronomy. In fact, the "North Star" (Polaris), is indeed a member of this class of variables. Henrietta Levitt, an astronomer who worked at Harvard in the early decades of the twentieth century, discovered the so-called "Period-Luminosity relationship." Hubble combined both redshift (to find velocities)

---

[2]These objects were called the "quasi-stellar radio sources" or "quasars" for short. It was found that these sources could not be stars in our galaxy but must be very far away, as far as any of the distant galaxies seen. We now think these objects are the very bright centers of some distant galaxies, where some sort of energetic action is occurring, most probably due to the presence of a supermassive black hole at the center of that galaxy.

and the idea of Cepheid variables (to calculate distance) to reach the now well-known relation $v = H_0 d$.

Nowadays, astronomers use three different methods to accurately measure distances to galaxies near to and far from the Earth and, as a result, to get the precise value of the Hubble constant. They use a cosmic distance ladder as shown below (Fig. 4.3).

As you can see from the this illustration, astronomers employ a basic tool that is well known in trigonometry, called parallax, to determine the distance to objects fairly close to the Earth. Once astronomers calibrate the Cepheids' true brightness, they can use them as cosmic yardsticks to measure distances to galaxies much farther away than they can with the parallax technique. For galaxies located even farther away from Earth, they employ another reliable benchmark, Type Ia supernovae, exploding stars

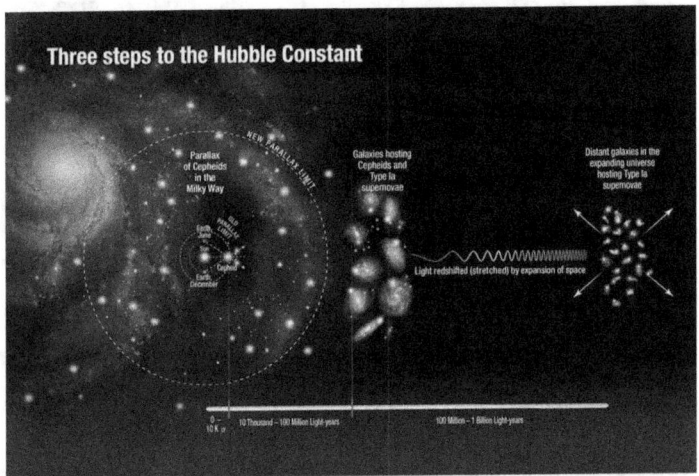

Fig. 4.3. *This illustration shows the three steps astronomers use to measure the universe's expansion rate to an unprecedented accuracy, reducing the total uncertainty to 2.4%. Image courtesy of NASA, ESA, A. Feild (STScI), and A. Riess (STScI/JHU).*

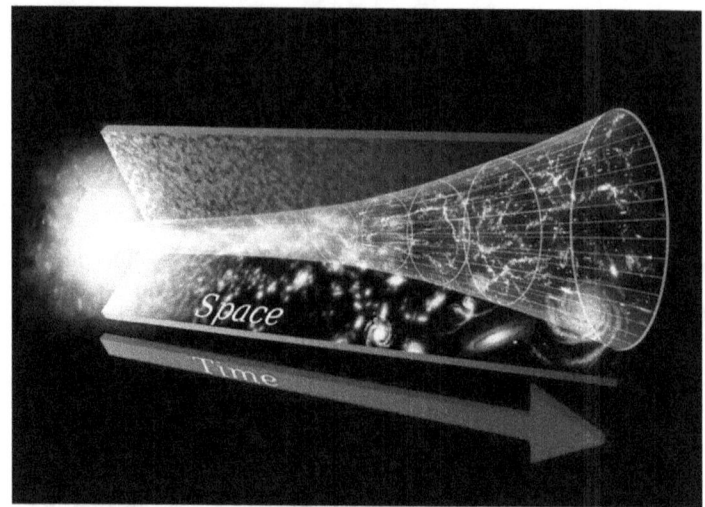

Fig. 4.4. *There is a large suite of scientific evidence that supports the picture of the expanding universe and the Big Bang. However, whether the universe is finite or infinite is not yet decided. Image courtesy of NASA/GSFC.*

that flash with the same amount of brightness. These can be seen from relatively longer distances as supernovae explosions are considered one of the greatest energy-releasing events in the universe. Astronomers compare the true and apparent brightness of distant supernovae to measure out to the distance where the expansion of the universe can be seen as shown in Fig. 4.3 (Three Steps to Hubble Constant, n.d.). They compare those distance measurements with how the light from the supernovae is stretched to longer wavelengths by the expansion of space (Fig. 4.4). They use these two values to calculate how fast the universe expands with time, called the Hubble constant.

Looking back in time

It is well known that when we look out at space, we are looking not just in space but back in time as well. The light we are seeing now from the farthest objects in the universe left those objects billions of years ago, so we see them as they appeared long ago. Thus, the answer to the question, "How far back we can see?" is not just related to space but to time as well as suggested in the general theory of relativity. Theoretically, we can see as far back in the universe as the point from where light is able to reach us, which is about 14 billion years ago since the birth of the universe. However, to compound this scenario, we should remember the universe is also expanding and stretching the fabric of space-time.

While the galaxies are flying away from us, the light particles, called photons, emitted by those galaxies are trying to reach us. These wavelengths of light get stretched out and redshifted to infrared, microwave, and even radio waves. Given time, the photons will be stretched so far that we may not be able to detect the galaxy at all. In the far future, all galaxies and radiation (energy) we see today from Earth may not be able to reach us and could become completely undetectable. It is possible that, in the distant future when our descendants look up at the sky, they will have no idea that there was ever a Big Bang or that there are other galaxies outside the Milky Way. So, we are in a race against distance and time. To look back, we employ magnificent observatories in space, such as the Hubble Space Telescope (HST) mentioned earlier (Fig. 4.5).

The HST has been operating since its launch in 1990—almost three decades now. Hubble's life in space has been longer than researchers had expected, and it has provided remarkable results with its deeper observation of the universe. Now, the aging Hubble will be replaced by the highly anticipated James Webb Space Telescope, which is scheduled to launch in March 2021. Webb is a multipurpose observatory that will allow astronomers to study some of the first stars and galaxies in the universe and even hunt for possible signs of

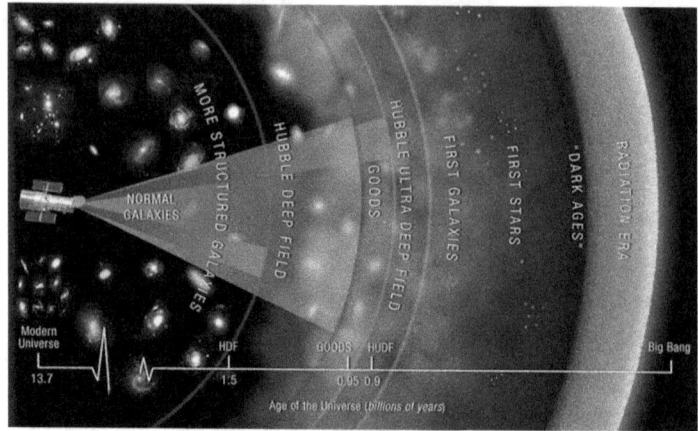

Fig. 4.5. *The deeper Hubble sees into space, the farther it gazes back in time. This chart illustrates the regions that have fallen under Hubble's eye. Image courtesy of http://hubblesite.org.*

life in the atmospheres of nearby alien planets (exoplanets—more on those in chapter 9).

As a comparison, it should be mentioned here that the primary mirror on the Webb Space Telescope is 21.3 ft (6.5 m) wide, compared to 7.8 ft (2.4 m) for Hubble's. Additionally, to get a sense of the orbital distances of these observatories, let us look at some more numbers. The Earth is 150 million km from the Sun, and the Moon orbits the Earth at a distance of approximately 384,500 km. HST orbits the Earth at an altitude of roughly 570 km. Webb, however, will not actually orbit the Earth; instead it will sit at the Earth-Sun Lagrange point, 1.5 million km away. Just to clarify, Lagrange points, named in honor of Italian-French mathematician Joseph-Louis Lagrange, are positions in space where the gravitational forces of a two-body system, like the Sun and the Earth, produce enhanced regions of attraction and repulsion. These locations can be used by spacecraft to reduce fuel consumption needed to remain in position.

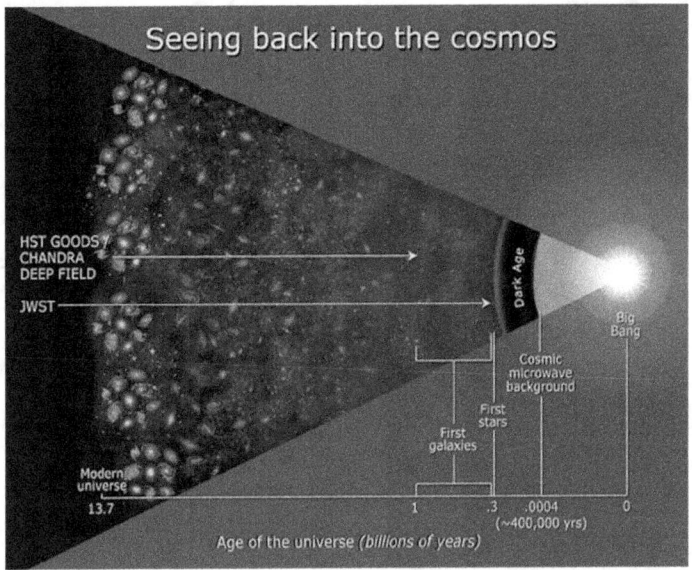

Fig. 4.6. *The ability to see back into the cosmos for different telescopes. Image courtesy of NASA and Ann Feild (STScI).*

The illustration above (Fig. 4.6) compares various telescopes and how far back they are able to see. If we say Hubble can see "toddler galaxies," then Webb will be able see "baby galaxies." One reason Webb will be able to see further, including first galaxies, is because it is an infrared telescope as opposed to HST, an optical telescope which operates mostly in visible spectrum. As we have seen in previous discussions, the light that originated from these early galaxies stretches as the universe expands and shifts the wavelength to longer ranges (remember the redshift), say from visible to infrared. As a result, distant objects are quite dim or even not visible for HST because that light reaches it as infrared light, although it started with shorter wavelengths. It is considered that powerful infrared telescopes, like Webb, are ideal for observing these early galaxies, and they can grant us a new vision of the cosmos.

For thousands of years, people have gazed in wonder at the world around them and asked the big questions: How did the universe come into existence? Where do human beings fit in this great cosmic scheme? Hubble's equation and his conclusion that the universe was expanding provided a new phase in our understanding of the cosmos. Furthermore, the new tools and technology we employ in learning about our own universe have contributed tremendously to our understanding. We have many more questions to answer, and it is arrogance on our part to claim science has all the answers, because science is a journey, not a destination. But people often seek comfort in ancient texts from various cultures to temporarily satisfy their curiosity. As a result, we frequently hear about the hidden wisdom in mythology sprinkled with mysticism, but labelling such myths as science is a dangerous proposition.

The myth of hidden scientific wisdom in eastern mysticism

Eastern philosophy and mysticism have had a long association with the West, conspicuously through modern physics, as a number of pioneering physicists were deeply interested in, and often influenced by, certain notions of the ancient Hindu scriptures, such as the Vedas and Upanishads. This interest was augmented and popularized by Fritjof Capra's book, *The Tao of Physics*, which was published in the seventies and still remains a source of great popular interest. However, the assertion by some Hindu nationalists that modern inventions, such as the internet or genetic engineering, were known and practiced by people in ancient times is not only a pure myth but an absurd thought that scientists never entertained.

The sense of oneness among the various elements of this universe and the role of consciousness as a component in explaining the mysteries of this physical world, as cited in ancient texts, seem to be an appealing source of scientific wisdom. In fact, the Nobel-Prize-winning physicist Eugene Wigner insisted that it was not possible to formulate the laws of quantum mechanics without reference to consciousness. Unfortunately, these ideas are speculative, and they suffer from the same sort of explanatory gap as classical theories. Therefore, it would be unscientific to jump to conclusions since science has not done its job completely as it continues its journey.

Physicists, including Bohr, Schrödinger, and others, have held the view that consciousness and the physical world are complementary aspects of the same reality. This view is far from the recent outcry of some Hindu nationalists that science is just figuring out what has already been described in the scriptures. So, when some Indian politicians, driven by nationalism and religious zeal, use the myth of hidden scientific wisdom in ancient scriptures to vindicate the so-called past glory, it should be seen as a tactic to fuel neo-nationalism or as a simple shameless display of ignorance in the public domain.

In my view, the inventions of modern science cannot be linked to any general metaphorical symbols displayed in the ancient texts. For example, Shiva's cosmic dance is a central metaphor in Capra's book, and a statue of Shiva adorns the CERN (the European Organization for Nuclear Research) compound as a gift from India, which enjoys observer state status at CERN. The parallel between Shiva's dance and the dance of subatomic particles that create and destroy the material world has been much discussed in several writings. However, this symbol does not equate the ancient wisdom and the modern scientific process of exploring matter, although at a metaphysical level an interpretation can be given about the creation and destruction of the universe. The Shiva statue is one of the many statues and art pieces at CERN, and one can find creation/destruction myths with the same common elements in many cultures. Objectively, this kind of symbolism is visible when we look at how planets are named after Roman Gods and Goddesses.

The Vedic teachings offer insights into the materialization and annihilation of the universe at a philosophical level. The cyclical nature of the universe and its age in the order of billions of years are quite appealing to scientific minds as they coincide with current scientific findings. While science provides insights into the workings of this universe, it also makes mistakes. Although, it is assumed that the scientific process corrects itself as it progresses. So, it would be unfitting to single out a scientific discovery or invention and claim it existed in ancient times since science is still progressing and may reject any existing scientific knowledge in the future as today's state-of-the-art technologies become future antiquity.

The proponents of this newfound ancient wisdom have one great chance to vindicate their argument. They can spell out one of the inventions hidden in ancient texts that science will come across in the future rather than wait for science to invent something and then predate it to the ancient references. They hesitate to do that but engage in juvenile elucidations of the scriptures to validate their claims by referring to these texts and linking them with existing scientific inventions.

The scientific awareness that built our modern civilization and the resulting tools we enjoy today are not the products or functions of any specific geographic area but are a reflection of our exploratory human nature propelled by curiosity. The philosophical dimensions of these ancient texts are worth considering as intellectual stimulants while we address the eternal questions in science. Nevertheless, the Vedic ideas or, for that matter, any ancient mythological or religious texts, do not offer the blueprint for existing or future inventions or problems.

If reading scriptures helps people to regard nature in a holistic manner, that is a noble cause. However, invoking divine intervention to get rain to fall in their own land is an action in vain. The most remarkable lesson to learn from the scriptures, obviously, is the message of the futility of war that can bring forth destruction and agony on humanity as detailed in the Indian classic Mahabharata. I have to echo what is said in Bhagavad Gita: "Better indeed is knowledge than mechanical practice."

We will continue our exploration of the universe that has been built on equations and telescopes which reflect our curiosity to know more and is based on rational thought and not on ancient or medieval stories. It is human imprudence to link the scientific process with myths. Johannes Kepler, the famous German astronomer, said: "If there is anything that can bind the mind of man to this dreary exile of our earthly home and can reconcile us with our fate so that one can enjoy living—then it is verily the enjoyment of the mathematical sciences and astronomy."

Bibliography

Batchelor, G. K. (2013). *An introduction to fluid dynamics.* Cambridge: Cambridge University Press.

Capra, F. (2010). *The tao of physics: An exploration of the parallels between modern physics and eastern mysticism.* Boston: Shambhala.

Commire, A., & Klezmer, D. *Women in world history: A biographical encyclopedia.* (n.d.). Gale Virtual Reference Library.

Doppler Effect. (n.d.). Retrieved from https://www.grc.nasa.gov/www/k-12/airplane/doppler.html

Ellis, G. F. R., Maartens, R., & MacCallum, M. A. H. (2012). *Relativistic cosmology.* Cambridge: Cambridge University Press.

Freedman, W. L., et al. (2001). Final results from the Hubble Space Telescope key project to measure the Hubble Constant. *The Astrophysical Journal, 553*(1):47–72.

Harrison, E. R. (2000). *Cosmology: The science of the universe.* Cambridge: Cambridge University Press.

Hubble, E. (1929). A relation between distance and radial velocity among extra-galactic nebulae. *Proceedings of the National Academy of Sciences of the United States of America, 15*(3):168–173.

Hubble, E. P. (1937). *The Observational approach to cosmology.* Oxford: Clarendon Press (printed by J. Johnson).

Hubble, E. P. (1936). *The realm of the nebulae.* New Haven: Yale University Press.

Hubble, E. P., & Humason M. L. (1931). The velocity-distance relation among extra-galactic nebulae. *The Astrophysical Journal, 74*:43–80.

HubbleSite: Out of the ordinary...out of this world. (n.d.). Retrieved from http://hubblesite.org/

James Webb Space Telescope, Goddard Space Flight Center. (n.d.). Retrieved from https://jwst.nasa.gov/about.html

Kepler's Discovery. (n.d.). Retrieved from http://www.keplersdiscovery.com/

Koestler, A. (2014). *The sleepwalkers: A history of man's changing vision of the universe.* London: Penguin Books.

Leavitt, Henrietta Swan (1868–1921). Retrieved 2019 from Encyclopedia.com: https://www.encyclopedia.com/women/encyclopedias-almanacs-transcripts-and-maps/leavitt-henrietta-swan-1868-1921

Lord Shiva Statue Unveiled. (June 24, 2004). Retrieved from https://cds.cern.ch/record/745737

Mascaro, J. *The Bhagavad Gita.* (1962). London: Penguin Books.

Mathew, S. (2014). *Essays on the frontiers of modern astrophysics and cosmology.* New York: Springer.

Morrow, A. (ed.). (2017). Three steps to measuring the Hubble Constant. Retrieved from https://www.nasa.gov/image-feature/goddard/2016/three-steps-to-measuring-the-hubble-constant

Three Steps to Hubble Constant (Artist's Illustration) (n.d.). Retrieved from http://hubblesite.org/image/3736/news

Voelkel, J. R. (1999). *Johannes Kepler and the new astronomy.* New York, NY: Oxford University Press.

Wainwright, J., & Ellis, G. F. R. (2005). *Dynamical systems in cosmology.* Cambridge, U.K.: Cambridge University Press.

Webb, S. (1999). *Measuring the universe: The cosmological distance ladder.* London: Springer.

CHAPTER 5

The equation that conceals a scientific tragedy

Available energy is the main object at stake in the struggle
for existence and the evolution of the world.
—*Ludwig Boltzmann* Quoted in D W Thompson's *On
Growth and Form (1917)*

Abstract

*The equation that represents Boltzmann's principle is appealing to
both scientists and philosophers. It connects the microscopic world to
the macroscopic world and opens up a completely new vista in modern physics. Its repercussions are still heard in various disciplines,
such as information, evolution, and atomic physics. However, this
equation, one of the most celebrated in science history, also conceals
a scientific tragedy.*

"The great tragedy of Science," wrote Thomas Henry Huxley, is "the slaying of a beautiful hypothesis by an ugly fact." Nevertheless, I do not want to begin this chapter with a scientific hypothesis or a fact but rather with a modest question.

What do Ludwig Beethoven and Ludwig Boltzmann have in common? Obviously, I am not asking about their first names or the identical letter with which their last names begin. While Beethoven constantly explored ways to break free of the styles that defined classical music, Boltzmann vigorously fought the orthodoxy and dogmatism of the scientific world. They both ended up in the Central Cemetery in Vienna, Austria.

Look at the top of Boltzmann's tombstone in the Vienna Central Cemetery (Fig. 5.1). There is an equation engraved on

Fig. 5.1. *Grave of Ludwig Boltzmann in Central Cemetery, Vienna, Austria. Image courtesy of Daderot, Wikimedia Commons.*

it which reads, S = k log W. This equation connects the entropy of a system (S) of particles to the number of possible "micro-states" (W) that particles can own. Entropy, the measure of disorder or randomness in a system, was a rather destructive idea for science, especially in the nineteenth century when stability and order were considered God-given assets (even now) that should prevail in a society. The constant k, later termed the *Boltzmann constant*, is approximately 1.3807×10^{-23} joule s per kelvin ($J \cdot K^{-1}$).

A bridge between microscopic and macroscopic worlds

In the eighteenth century, Antoine Lavoisier, the founder of modern chemistry, attempted to replace the ancient idea that everything was made up of earth, air, water, and fire. He considered that heat was an invisible substance similar to ether that filled the spaces through which light waves traveled. It turned out that both of these ideas were wrong, and we do not need those extra assumptions to explain the flow of heat or the propagation of light waves. In the case of heat, this unseen material was called "caloric fluid," and it was believed that it could be transferred between objects but could be neither created nor destroyed.

Additionally, it was thought that heating up an object meant that caloric fluid would flow into it and, surprisingly, this idea was used to explain the expansion of objects when heated (Fig. 5.2). However, this kind of knowledge did not last long. For example, the idea of caloric fluid could not explain how heat could emanate from a cold piece of wood once it was set on fire.

71

Fig. 5.2. *In 1787, Lavoisier, the founder of modern chemistry, called heat "the caloric fluid," from the Greek word for heat. Heating or burning meant the flow of this invisible caloric fluid. Image courtesy of Pexels from https://www.pexels.com/photo/blaze-burn-burning-burnt-97492/.*

Where did the caloric fluid come from? If it had been in the wood in the first place, the wood should have been hot all along.

This is where Boltzmann's contributions to modern physics become so important, and they go beyond the entropy equation mentioned above. He extended the work of Scottish physicist James Clerk Maxwell to establish a kinetic-molecular theory with the following assumptions for a system of gases:

- the gas is composed of a large number of identical molecules moving in random directions, separated by distances that are large compared with their size;
- the molecules undergo perfectly elastic collisions (no energy loss) with each other and with the walls of the container, but otherwise do not interact;

- the transfer of kinetic energy between molecules is heat (heat is no longer the invisible caloric fluid).

These simple and logical assumptions may seem very familiar now. However, these ideas were so remarkable at the time, in the sense that they provided new meanings for existing ideas like heat and temperature. Consequently, the caloric theory was abandoned in the nineteenth century and replaced with the kinetic-molecular theory. Soon, the theory paved the way for one of the most important ideas in modern science that established the fact that all matter is made up of molecules and atoms in constant motion.

According to kinetic theory, heat is no longer an invisible fluid as was thought by a previous generation of scientists. In fact, molecular motion is the cause of temperature increases in a body, and when the molecules move faster, the temperature is higher. As discussed in chapter 3, although Newton put together a theory on how the world worked, his theory was now challenged by new discoveries and ideas such as the motion of atoms and molecules. In fact, physics was about to undergo massive changes based on these new findings. The discovery of molecular motion and its connection to temperature and heat was a huge paradigm shift in the world of physics.

In fact, during these years (the second half of the nineteenth century), enormous changes were visible not only in physics but also in other disciplines and in socio-political environments. We need to remember that it was a time of upheaval. Cracks were beginning to appear, not just in the Austro-Hungarian empire where Boltzmann lived but were also emerging in the traditional scientific approach, and revolutionary ideas were floating around. Charles Darwin published *Origin of Species* in 1859 and Georg Cantor, in 1874, provided the proof for a distasteful idea in mathematics called *infinity*. All these revolutionary ideas had the potential to create a new world order founded on new social and scientific constructs. It should be mentioned here that Boltzmann was considered one of the early physicists who supported Darwin's evolutionary theory.

Although he made several other contributions to modern physics, Boltzmann's greatest scientific achievement can be summarized by the equation $S = k \log W$. So, what makes this equation so significant and, at the same time, a tragic symbol of science? To understand this equation, let us look at each variable separately.

In 1865, Rudolf Clausius, a German mathematical physicist, introduced the concept of entropy (S) to science. However, this idea was not purely philosophical or theoretical, as we might think. In those days, while designing steam engines, engineers realized that energy was always lost when converted to work. In other words, no machine could be 100% efficient. In a practical sense, the term "entropy," from the Greek word for transformation, represents this energy loss.

The significance of Boltzmann's work lies in the fact that he expanded the idea to a much wider and deeper arena in physics. In his 1877 paper, titled "Further Investigations on the Thermal Equilibrium of Gas Molecules," Boltzmann stated that entropy is a measure of the disorder of the state of a physical system, and that nature can be described by statistics. The use of statistical probability in analyzing the fundamental properties of matter is what makes Boltzmann's principle, represented by $S = k \log W$, one of the most intriguing equations in physics.

Now, regarding the letter W in the equation, it represents the number of microstates for a configuration of a system. It is the number of ways in which a configuration in a system of molecules or atoms can be put together. For example, if there are four molecules and only two possible orientations for each molecule to be arranged, then $W = 2^4 = 16$. In fact, W directly gives you the probability of the macrostate and log W is then the logarithm of this probability.

Finally, the constant k, termed the *Boltzmann constant*, is approximately 1.3807×10^{-23} joules per kelvin ($J \cdot K^{-1}$). It represents the energy in a gas molecule and is directly proportional to the absolute temperature. As the temperature increases, the kinetic energy per molecule increases.

The notion that nature can be described by statistical probability is the core of this principle, which is capable of transforming our world view. Boltzmann's statistical conception of mechanics and thermodynamics provided a framework with which the microscopic and macroscopic worlds could be connected. So, here is the revolutionary idea brought by Boltzmann's equation—the macroscopic properties of a system such as heat or work can be derived from the individual properties of atoms and molecules, such as temperature and pressure. It reveals that the interactions of the microscopic world determine the features of our real world. Furthermore, these interactions are random, and all we can do is sum them up approximately and predict a probable state for our world. This is rather shocking, especially since it has long been thought that the underlying mechanism that drives our universe is certainty, and order is a prime feature of that determinism.

Boltzmann's attempt to connect the microscopic world to the macroscopic world had indeed attracted severe opposition from the scientific community. As described above, microstates, W in the equation, refer to the state of the atoms and molecules that constitute a system. The trouble was that no one could see the atoms. It was a scientific heresy and, as one of Boltzmann's opponents, Ernst Mach, remarked, "Atoms? Have you seen one yet?" Moreover, according to Boltzmann, the entropy or disorder was an emergent phenomenon of the interacting particles at a microscopic level. For that matter, order could also be considered as an emergent phenomenon arising from within the system and not a God-given virtue from above as one would expect.

Philosophy and tragedy

Ludwig Eduard Boltzmann was born February 20, 1844, in Vienna, to an affluent family and received his preliminary education from a private tutor in the house of his parents. In 1863, he entered the University of Vienna to study physics and received his doctorate in 1866 for a thesis on the kinetic theory of gases supervised by Josef Stefan. At the age of twenty-five, he became a full professor of mathematical physics at the University of Graz in the province of Styria, and during this period, he devoted much of his time to exploring the statistical concept of nature. Boltzmann devoted the second part of his life mostly to philosophy (Fig. 5.3).

In the book, *Physics and Philosophy*, Paul Feyerabend (2015) remarked about Boltzmann: "In his realization of the hypothetical character of all our knowledge, Boltzmann was far ahead of his time and perhaps even our time." The concept of matter and its fundamental constituents has undergone various refinements in the history of science. In the nineteenth century, the scientific community was skeptical about such changes. As a result, Boltzmann faced severe criticism from science traditionalists for his revolutionary work.

Fig. 5.3. *Boltzmann, physicist and philosopher. Image courtesy of Wikimedia Commons: https://commons.wikimedia.org/wiki/ File:Boltzmann_age31.jpg.*

Boltzmann's grandson, Dieter Flamm, wrote, "Boltzmann was a martyr to his ideas." Unfortunately, the objections made to him were not scientific discussions but mere dogmatisms. According to Max Planck, who was initially opposed to Boltzmann but later converted and used his approach, "against the authority of men like Ostwald, Helm, and Mach there was not much that could be done."

We can also see Boltzmann's principle through the lens of the socio-political systems of those days. His philosophy was rooted in atomism and against the positivism of the Christian world where nature was seen as the product of a transcendent creator and was therefore considered rational. The purpose and order provided by Aristotelian notions were more suitable to the positivists. We have to see the Boltzmann equation carefully against the backdrop of those socio-political systems and similar orthodox systems that existed in the scientific approach where proof meant observable evidence. Yet, he insisted on the necessity of accepting the reality of the external world, and this is often labelled in philosophy as realism, and later materialism.

Readers can get a sense of Boltzmann's approach to science and philosophy when he described his appreciation of nature and his philosophy in these words:

> I wept when I saw the color of the sea—how can a mere color make one cry? Or moonlight, or the luminescence of the sea in a pitch black night?... But if there is one thing which is more worthy of our admiration than natural beauty, it is the art of men who have conquered this never-ending sea so fully in a struggle that has been going since the time of the Phoenicians. (Quoted in, George Greenstein, 'The Bulldog: A Profile of Ludwig Boltzmann', The American Scholar, 1991)

Boltzmann tried to link his natural philosophy with realism and took physics and philosophy in a totally new direction that was unheard of before.

Boltzmann strongly rejected the notion of absolute certainty but deeply favored an approach based on probability in explaining natural phenomena. He believed that stability and order could be appealing to those who provided them but were not appealing for those who had to accept them. He was quoted as saying, "The stars bend like slaves, to laws not decreed for them by human intelligence, but gleaned from them."

Although Boltzmann held uniquely strong convictions in science and philosophy and stood against the scientific orthodoxy of his day, his opponents criticized him relentlessly. His theories were not given the respect they deserved by his rivals, and he was haunted by his own struggles with physical and mental health. Boltzmann took his own life in 1906 during a holiday in Italy. His theory received widespread acceptance only after his death when Max Planck and Albert Einstein realized the significance of his work.

Significance of Boltzmann's work

It is well known that Max Planck introduced quantum mechanics in the 1900s based on the concept of quantization of energy, which essentially means energy comes in discreet packets, not as a continuous wave as once thought. Boltzmann's statistical view played a huge role in Planck's formulation of quantum theory (more on Planck in chapter 7). Initially, Planck was not ready to accept Boltzmann's statistical approach but did accept it later on. If we look at the scientific literature carefully, we can see that the idea of quantization of energy originated from the works of Boltzmann.

Although Boltzmann never used the term energy quantization in his work, we can trace this idea in his papers decades before Planck formally introduced the term. In computing statistical probabilities, Boltzmann considered the energy of a system as a combination of small packets—the very same idea of quantized energy. Again, it should be emphasized here that his approach was to bridge the microscopic properties to the real world, and he continued in this way until his death.

Boltzmann's opposition to the strictly deterministic view of physics is evident in both the science and philosophy that he stood for. His statistical interpretation of entropy was a major departure from the classical view about the fundamental laws of physics. In addition to the molecular theory he is known for, his contributions to statistical mechanics were instrumental in shaping quantum mechanics. Although various methods exist now in statistical mechanics, Boltzmann's approach is still an important one and is applicable to various purposes.

The significance of atomism, as advocated by Boltzmann, is evident in our contemporary world. We live in a world built on the economics of atoms and electrons. Boltzmann's atomistic view and statistical approach paved the way for parts of modern physics, particularly quantum mechanics. Many modern subjects, like biochemistry, rely on thermodynamics and statistical mechanics. As you may recall, we have seen in chapter 1 how the concept of entropy, used in information science by Shannon, revolutionized the field. Similarly, in many areas where probabilistic outcomes dictate the result, whether in computer science, economics, or financial systems, Boltzmann's approach remains relevant and could be explored further.

In his popular lecture on the second law of thermodynamics, Boltzmann made the following statement about evolution and biology (Cercignani, 1998):

> The general struggle for existence of living beings is therefore not a fight for the elements—the elements of all organisms are available in abundance in air, water and soil—nor for energy, which is plentiful in the form of heat, unfortunately untransformably, in every body. Rather, it is a struggle for entropy that becomes available through the flow of energy from

the hot Sun to the cold Earth. To make the fullest use of this energy, the plants spread out the immeasurable areas of their leaves and harness the Sun's energy by a process that is still unexplored, before it sinks down to the temperature level of the Earth, to drive chemical syntheses of which one has no inkling as yet in our laboratories. The products of this chemical kitchen are the subject of the struggle in the animal world.

It should be mentioned here that revolutionary theories, like evolution and entropy, challenged many people's views of the so-called "natural" order. Both evolution and entropy moved beyond their original fields, shaped our perspective, and formed a new world order (Fig. 5.4).

At this juncture, it is also worth thinking about a few questions in regards to biological life and entropy, even though, at present, we may not be able to answer the following questions: Can life originate from the laws of physics? Can inanimate

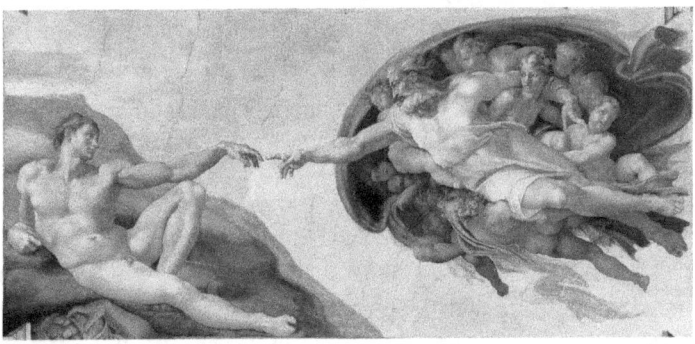

Fig. 5.4. *Can life originate from the laws of physics? If so, the second law of thermodynamics and the concept of entropy are key to that revolution. Image in the public domain courtesy of Wikimedia Commons.*

matter inexorably acquire the key physical characteristics that we associate with life? Can the principles of physics that govern everything from atomic energy to gravitational waves be applied here?

We know that the essential difference between living things and inanimate matter is the absorption and use of energy; living things can do that in a much more efficient way. Another feature that distinguishes living organisms from non-living ones is self-replication. A simple physical intuition dictates that this process, common to every species of living thing, must invariably be fueled by the production of entropy. It is astounding to note that the concept of entropy, deeply explored by Boltzmann, could be relevant in various fields even now. The statistical physics of self-replication might be at work as indicated in a recent work by Jeremy England from MIT (2013). His mathematical formula, based on the concept of entropy, indicates an increase in energy dissipation when a group of atoms is driven by an external source of energy and surrounded by a heat bath (like the ocean or atmosphere). This could be interpreted as an indication that mimicking nature's ability to self-replicate is within the realm of practical science and will be another milestone in understanding biological mechanisms in a deeper way. Again, entropy could play a role in organizing matter in such a way that it could attain the key physical attributes that we associate with life.

Clearly, we are not done with entropy, and its implications will be felt in the future more than now—an idea that originated from work with steam engines and that is enriched by Boltzmann, both philosophically and scientifically. Boltzmann's insight to connect fundamental physics to philosophy and other disciplines was a huge accomplishment. Ironically, Boltzmann's powerful philosophy and equation point to the fact that entropy is irreversible and will lead only to decay with no scope for eternity, but certainly his equation will remain eternal.

Bibliography

Blackmore, J. (1995). *Ludwig Boltzmann: His later life and philosophy, 1900–1906*, Dordrecht: Kluwer.

Broda, E. (1983). *Ludwig Boltzmann: Man, Physicist, Philosopher*. Woodbridge, CT: Ox Bow Press.

Cercignani, C. (1998). *Ludwig Boltzmann: The man who trusted atoms*. New York: Oxford University Press.

Cropper, William H. (2001). *Great physicists: The life and times of leading physicists from Galileo to Hawking*. New York: Oxford University Press.

England, J. L. (2013). Statistical Physics of self-replication. *Journal of Chemical Physics, 139*(12). Retrieved from https://aip.scitation.org/doi/10.1063/1.4818538

Feyerabend, P. K., Gattei, S., & Agassi, J. (eds.). (2015). *Physics and philosophy: Philosophical papers*. Cambridge: Cambridge University Press.

Flamm, D. (1973). Life and personality of Ludwig Boltzmann. In Cohen, E. G. D., & Thirring, W. (eds.), The Boltzmann Equation. *Acta Physica Austriaca* (Supplementum X Proceedings of the International Symposium "100 Years Boltzmann Equation" in Vienna, 4th–8th September 1972). Vienna: Springer.

Flamm, D. (1983). Ludwig Boltzmann and his influence on science. *Studies in History and Philosophy of Science, 14*(4), 255–278.

Garber, E., Brush, S. G., & Everitt, C. W. F. (eds.). (1995). *Maxwell on heat and statistical mechanics*. Bethlehem, PA: Lehigh University Press.

Gibbs, J. W. (1902). *Elementary principles in statistical mechanics*. New York: Scribner.

Greenstein, G. (1991). SCIENCE: The Bulldog: A Profile of Ludwig Boltzmann. *The American Scholar*, 60(1), 97–105. Retrieved from http://www.jstor.org/stable/41211873

Khinchin, A. I. & Gamow, G. (1949). *Mathematical foundations of statistical mechanics* (pp. 74–75). New York: Dover Publications Inc.

Klein, M. J. (1973). The development of Boltzmann's statistical ideas. In Cohen, E. G. D., & Thirring, W. (eds.). *The Boltzmann Equation* (pp. 53–106). Vienna: Springer.

Mayer, J. E. & Mayer, M. G. (1977). *Statistical mechanics*. New York: John Wiley & Sons.

Von Plato, J. (1994). *Creating modern probability*. Cambridge: Cambridge University Press.

Thompson, D. W. (1917). *On growth and form*. Cambridge: University Press.

CHAPTER 6

The equation that stole eternity

The law that entropy always increases [the second law of
thermodynamics] holds, I think, the supreme position
among the laws of Nature.
—Sir Arthur Stanley Eddington, The Nature of the
Physical World (1928)

Abstract

*The second law of thermodynamics was first proposed by two scien-
tists, Rudolf Clausius and William Thomson (Lord Kelvin), using
different examples. It has also been claimed that the French physicist
Sadi Carnot discovered the same idea some twenty-five years earlier. The second law of thermodynamics states that there is a natural
tendency in any isolated system to degenerate into a more disordered
state. In other words, our universe as an isolated system is bound
for more disorder and, eventually, destruction. It implies that the*

universe will end in a "heat death" in which everything is at the same temperature. This is the ultimate level of disorder. If everything is at the same temperature, no work can be done, and all the energy will end up as the random motion of atoms and molecules.

Before we begin to talk about the second law of thermodynamics or entropy, let us begin with a closely associated idea. The concept of eternity has a deceptive beauty attached to it, and it varies in its appeal depending on the consumer of this product. Ironically, it works with rich and poor alike. Often, it is offered as a panacea for our earthly sufferings, from hunger to severe depression. Furthermore, it is interpreted as an extension of this life, and if any one questions that possibility, it is presented as a new phase in our life. Sometimes, people sprinkle philosophical thoughts over it so it seems a much more complicated idea and suggest we should not try to understand it, but we should get ready for it. In modern times, the dream of eternal life has been used to nurture religious extremists and, in theocratic societies, it is an instrument more powerful than the military system.

Around 1850, while studying basic thermodynamic systems, physicists Rudolf Clausius and William Thomson (Lord Kelvin) discovered that heat does not spontaneously flow from a colder body to a hotter body. In fact, this realization became the basis for the second law in its various forms. In the previous chapter, we discussed the statistical nature of entropy brought forth by Ludwig Boltzmann, but here we are going to focus on thermodynamic entropy. Also, remember how we connected the idea of information entropy in chapter 1 using Shannon Entropy. The laws of thermodynamics define and relate fundamental properties (temperature, energy, and entropy) of thermodynamic systems similar to Newton's laws that described mechanical systems. In classical thermodynamics, the entropy of a system is the ratio of heat content to temperature. In simple mathematical form, for a reversible process, the equation is written in the following form:

$$\Delta S = \frac{Q}{T}$$

S – entropy of the system
Q – the heat content of the system
T – the temperature of the system

This means that, if a reversible process occurs, there is no net change in entropy. However, in an irreversible process (most of the natural processes), entropy always increases, so the change in entropy is positive. The total entropy of the universe is constantly increasing as represented mathematically:

$$\Delta S \geq 0$$

Additionally, the second law of thermodynamics can be stated in several ways:

- It is not possible to convert one form of energy to another with 100% efficiency.
- Heat will not flow spontaneously from a cold body to a hot body.
- The universe, as a closed system, is leading to a heat death.
- Left to themselves, all things tend to equalize their temperatures. In an isolated system, eventually everything will be the same temperature, and heat will no longer be able to flow.

As we discussed in previous chapter, entropy can be interpreted as randomness or disorder of the system. It also represents

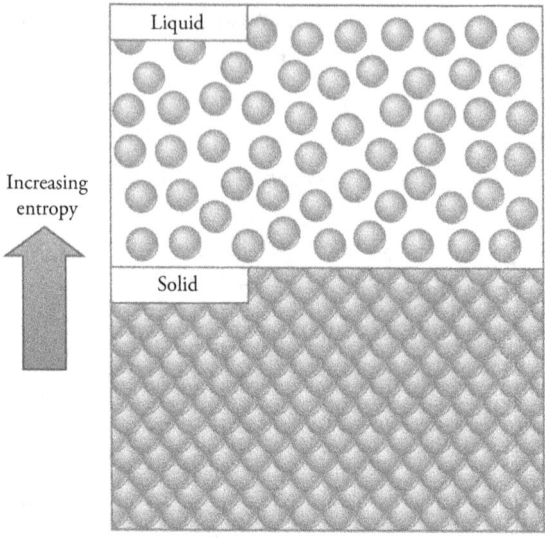

Fig. 6.1. *The basic idea of entropy is shown in this illustration. When the disorder of the molecules increases, entropy also increases, which means less useful energy.*

the unavailability of energy to do work (Fig. 6.1). Energy, in physics, is the capacity for doing work. It may exist in various forms such as potential, kinetic, thermal, electrical, chemical, nuclear, or other forms.

In order to better understand the concept of entropy, we should also look at the internal energy of a system. The internal energy (often called thermal energy) is the energy due to the kinetic and potential energies associated with the random motions of all the particles (molecules and atoms) that make them up. The kinetic energy is, of course, due to the motion of the particles. The potential energy of the molecules arises from the attractive and repulsive forces acting between molecules that result in molecular bonds. The potential energy depends on the strength of the bonds; strong bonds have low energy and weak ones have high energy. Even among physicists, internal energy and thermal energy are often used in a vague or interchangeable fashion. However, to make this distinction clear, we can consider internal energy as the sum of the kinetic energies and potential energies for the system and thermal energy as a general term that is closely related to internal energy.

When heat (energy) is supplied to a thermodynamic system, part of that is used to increase the internal energy, and the rest is for work. One thing the second law explains is that it is impossible to convert heat energy to mechanical energy with 100% efficiency. If 100 J (joule is the unit of energy and work) of energy is given to a system to do the work, there is always some leftover heat in the system that cannot be used to do any additional work. Moreover, we know that any movable parts produce friction that converts mechanical energy to heat which is generally unusable. Indeed, claims for perpetual machines are rejected by the U.S. Patent Office due to this constraint set by the second law of thermodynamics.

Although it's difficult to measure the total entropy of a system, it's generally fairly easy to measure changes in entropy (ΔS). For a thermodynamic system involved in a heat transfer of size Q at a temperature T, a change in entropy can be measured by the above equation ($\Delta S = Q/T$). This law holds a supreme position in the hierarchical order of natural laws. As the physicist Arthur Eddington wrote in The Nature of the Physical World (1928):

> The law that entropy always increases holds, I think,
> the supreme position among the laws of Nature. If
> someone points out to you that your pet theory of the
> universe is in disagreement with Maxwell's equations—
> then so much the worse for Maxwell's equations. If it is

found to be contradicted by observation—well, these experimentalists do bungle things sometimes. But if your theory is found to be against the second law of thermodynamics, I can give you no hope; there is nothing for it but to collapse in deepest humiliation.

The unavailability of energy to drive a system due to the inevitable energy loss is what Clausius, in 1865, called entropy (the word originates from the Greek word *entrope* which means change) (Fig. 6.2). Scientists agree that the second law of thermodynamics is, in fact, a supreme law that governs the evolution of this universe and our place in it. However, it is to be emphasized here that the most important aspect of this law is that it applies

Fig. 6.2. *These ice floes melt during the Arctic summer. Some of them refreeze in the winter, but the second law of thermodynamics predicts that it would be extremely unlikely for the water molecules contained in these particular floes to reform the distinctive alligator-like shape they formed when this picture was taken in the summer of 2009. Image courtesy of Patrick Kelley, U.S. Coast Guard, U.S. Geological Survey.*

to isolated systems. These systems do not exchange energy with their environment or with another system.

As mentioned earlier, the most important aspect of this law is that, in its pure form, it applies to closed systems, ones that do not exchange energy with another system. Our universe is considered a closed system, and that means it should come to an end in the far future. Belief in the philosophical eternity promised in scriptures was a way of life for many early thinkers and scientists. Nevertheless, many kings and monarchs, who had traditionally claimed to rule by divine law, deployed that tool to distract the common man from his daily hardship. So, any attempt to disturb the order and stability was against the divine law. I do not want to sound pessimistic but, unfortunately, the second law of thermodynamics will disappoint all of us: there is no eternity, and nothing is free. As Stephen Hawking remarked, "We could call order by the name of God, but it would be an impersonal God. There's not much personal about the laws of physics."

While physicists agree with the immutability of this law, many attempts have been made to apply the second law of thermodynamics in other domains of human endeavors, including social and evolutionary fields.

Social entropy and the second law of thermodynamics

The above discussion has shown how the original idea of entropy is linked to thermodynamics, the study of heat and work. Now we could ask the question, "Why has this law become so charming and precious to philosophers, sociologists, and physicists alike?"

Steven Pinker, a cognitive psychologist at Harvard University, writes (2017):

> To start with, the second law implies that *misfortune may be no one's fault.* The biggest breakthrough of the scientific revolution was to nullify the intuition that the universe is saturated with purpose: that everything happens for a reason. In this primitive understanding, when bad things happen—accidents, disease, famine—someone or something must have *wanted* them to happen.

Clearly, the second law of thermodynamics has found its place in the social and psychological domains, once again underscoring the importance of this law.

The application of scientific theories in the social realm can be open to never-ending debates and often leads to dangerous propositions, as we learned from history. Another veiled hazard when scientific theories are extended vis-à-vis social states is the likelihood of providing plenty of room for misappropriation and misrepresentation of the original ideas.

The best known instance of a scientific theory used to explain social phenomena is social Darwinism, a term coined in the late nineteenth century to describe the idea that humans, like animals and plants, compete in a "struggle for existence" in which natural selection results in "survival of the fittest." Since its inception, many have used this theory to justify their own view of social, political, or economic policies or even to justify imperialism or racism. Ironically, the phrase "survival of the fittest" associated with the theory of evolution was not originally introduced by Darwin but by the sociologist and philosopher Herbert Spencer. In the very same way, "struggle for existence" had its origins in the writings of political economist Thomas Malthus, who had promoted the idea that poverty and famine were natural outcomes of population growth and food supply.

As we have learned from our own experience on this planet over the last centuries, the idea of social Darwinism and its predicted outcomes do not stand a chance against human ingenuity. We can overcome the constraints set by natural selection and survive as a society. Spencer's argument that competition and social evolution would automatically produce a level of prosperity and personal liberty unparalleled in human history is flawed. These types of arguments downplay our human abilities to overcome poverty and warfare with social reforms and proper policies. Moreover, if social Darwinism, as envisioned by the above thinkers, was working in our world affairs, there should be less population growth in parts of the world where fewer resources and more competition for them exist, but we know it is the other way around.

Although social Darwinism was entertained in the past by many, it has been fading from the socio-political world for a long time. The only recent flare-up was when then President Obama invoked this term to describe a House Republican budget plan as "thinly veiled social Darwinism" (2012). So, scientific theories applied to social norms can be metaphors at best but not a precise way of measuring the socio-economic status of a society.

Not surprisingly, entropy has also been associated with social theories. Similar to social Darwinism, it is known as social entropy. A literature review shows a resurgence in the use of social entropy, and we must inquire whether it is heading down the same dangerous path set by social Darwinism.

Although Peter Tait, a Scottish mathematical physicist, is credited as having given us the earliest English use of the term "social entropy" in 1874, it was Henry Adams, the American historian and philosopher, who popularized the phrase. Adams used the second law of thermodynamics, or, for that matter, the concept of entropy, to argue that there is a tendency in the histories of whole nations or civilizations that is characterized by a running down of human energy, a diminished capacity for meeting the problems of that nation or civilization. He stated that all energy dissipates, order becomes disorder, and the Earth eventually becomes uninhabitable. In other words, he thought the social chaos around him was merely the inevitable outcome of this mighty law.

Again, Steven Pinker writes (2017), "Poverty, too, needs no explanation. In a world governed by entropy and evolution, it is the default state of humankind." Nevertheless, this whole notion, despite its good intention to direct the debate to the root cause of poverty, lacks a proper scientific footing. We can see that hunger and poverty exist in this world regardless of the availability of resources. Imagine a world in which a bunch of people are the only inhabitants; that means the entire resources of the planet are available for them. However, we can guarantee that hunger will be their most probable condition. The resources must flow within different parts of the systems for them to work, and the fundamental mechanism that makes that possible is the default state of humankind, that is, human ingenuity.

Similarly, we can see poor and poverty-stricken populations surrounded by wealthy segments in many parts of the world where plenty of resources already exist. Therefore, hunger and poverty are not the default state; rather the default state is the human ingenuity by which we overcome these issues. That is what we see in this world. Thus, poverty can exist whether resources are available or not, but we know the world has gotten better in dealing with poverty; it cannot be the default state of humankind. We also know that our ancestors, in fewer numbers, occupied this planet and definitely encountered hunger, but something drove them to surpass those constraints.

The trouble begins when we consider entropy as the manifestation of thermodynamic entropy. First, the second law of thermodynamics says the entropy of a closed system increases. Earth receives energy from the Sun and radiates a part of that energy back which makes Earth far less than a closed system (let alone humans). Nature is capable of generating order out of disorder on a local level without violating the second law of thermodynamics, and human ingenuity can sustain that process. On this planet, entropy can be decreased and order can be maintained while society advances. The crude application of the second law of thermodynamics to Earth or humans does not yield any logical results.

We could also analyze entropy on the molecular level. We know the faster the molecules move, the more the entropy of the system. In essence, entropy represents a dynamic system where the movements of the components are possible. So, it is not as bad as we perceive it to be, at least in a social sense. In the absence of entropy, we would have a stagnant society where no action would take place, and we would end up in a useless state of equilibrium.

As a physicist, I have no doubt about the power of the second law of thermodynamics—the entropy of an isolated system remains constant or increases. Einstein remained convinced throughout his life that thermodynamics was the only universal physical theory that would never be refuted. At a deeper level, it is hard to give a simple description of this fact as we continue to explore our universe. Our current understanding about the universe as a closed system could change in the future, and if that happens, we should be prepared to offer a better explanation for entropy and disorder, which should be different than what we say now. Not all cosmologists agree on the heat death predicted for our universe by the second law, although the law seems to drive the universe in only one direction.

Boltzmann was quoted in D'Arcy Wentworth Thompson's *On Growth and Form* (1942): "Available energy is the main object at stake in the struggle for existence and the evolution of the word." Similar to other systems, in the case of living organisms, some energy is lost in the process of energy transfer (accepting the fact that living organisms are not closed systems), and the entropy increases. The flow of energy is the mechanism that maintains order in cells and, as a result, life. In the absence of this transfer, entropy has the upper hand, and death is warranted. So, it is quite natural to think that we can see evolution through the lens of entropy.

Evolution and the second law of thermodynamics

The claim that evolution violates the second law of thermodynamics has often been misinterpreted. The rationale behind this argument is that entropy represents disorder or randomness, but evolution indicates progress from disorder to order as complex and organized life forms emerge from less complex ones. Organization and complexity increase during evolution.

As mentioned earlier, the second law of thermodynamics predicts an increase in entropy for isolated systems. An isolated system is one that does not exchange energy with its surroundings. Therefore, much care is needed when we apply this law to systems which are not closed, and the systems that we are referring to in regards to evolution are not isolated systems.

For example, the Earth continuously receives energy from the Sun and cannot be considered an isolated system. Life is not a closed system as any form of life is in continuous interaction with other systems, including its surroundings (Fig. 6.3). The maintenance of

Fig. 6.3. *Does evolution violate the second law of thermodynamics? If one wants to look at evolution and entropy, the role of the surroundings or the environment is a key factor. Evolution doesn't take place in a closed system. Image is in the public domain via Wikimedia Commons.*

order with the flow of energy to diminish the impact of disorder is a valid process for living systems—they create order from disorder. It is common in non-living systems as well, as you will see in chapter 10. In spirit, we cannot apply this law holistically for a system like Earth, or life itself, without considering its environment or surroundings; evolution does not take place in a closed system.

Furthermore, evolution requires the existence of outside forces, such as natural selection, and is irreversible, but the entropy increase must happen, not just in the system but in the system and its surroundings. A remark by French biochemist Jacques Monod, who shared the 1965 Nobel Prize for medicine, is appropriate here. In his 1972 work, *Chance and necessity: An essay on the natural philosophy of modern biology*, he wrote, "Evolution in the biosphere is therefore a necessarily irreversible process defining a direction in time; a direction which is the same as that enjoined by the law of increasing entropy, that is to say, the second law of thermodynamics. This is far more than a mere comparison: the second law is founded upon considerations *identical* to those which establish the irreversibility of evolution. Indeed, it is legitimate to view the irreversibility of evolution as an expression of the second law in the biosphere."

To encompass the system and its surroundings, we can modify the second law mathematically as follows:

$$\Delta S_{universe} = \Delta S_{system} + \Delta S_{surroundings} > 0$$

This means an increase in order created locally within a system is a possibility, but it can increase the disorder in the surroundings. We cannot alter the fact that overall entropy increases as does, obviously, disorder. The second law still wins.

Biologists agree that age changes can occur in only two fundamental ways: either by a purposeful program driven by genes or by random or accidental events. The application of the second law for a biological system may not be appropriate. However, there are strong arguments to support the application of this law in the process of aging. Leonard Hayflick (2007), with the Department of Anatomy, University of California, San Francisco, emphasizes that the common denominator that underlies all modern theories of biological aging is change in molecular structure and, hence, function. These changes are the result of entropic changes. This claim is now supported by the recent reinterpretation of the second law of thermodynamics where the belief that it only applies to closed systems has been overturned. Hayflick is referring to other works that support the belief that age changes are characterized by increasing entropy, which results in the random loss of molecular fidelity and accumulates to slowly overwhelm maintenance systems. All the discussions underscore the fact that energy and entropy are very valuable concepts in understanding the workings of this universe for living and non-living things alike, and they are going to remain in our science and discussions for a long time.

It is possible that, in the future, we will rewrite the immutable laws of nature in light of fresh evidence. But, for now, we are stuck with the second law of thermodynamics and its impact. We could consider entropy as a tendency for energy to become unavailable for useful purposes regardless of whether the system is open or closed. In the paper, Biological aging is no longer an unsolved problem, Leonard Hayflick (2007) writes, "The prevention of chemical bond breakage, among other structural changes, is absolutely essential for life. Through evolution, natural selection has favored energy states capable of maintaining fidelity in most molecules until reproductive maturation, after which there is no species survival value for those energy states to be maintained indefinitely." The strength of this law in determining the aging process of humans and the universe is what makes it a supreme law of physics.

What does entropy teach us? We can interpret it in myriad ways including descriptive or computable ways. In classical mechanics, the change in entropy represents the amount of energy input to the system which is not available for the mechanical work done by the system. In statistical mechanics, entropy becomes a function of statistical probability as discussed in the Shannon and Boltzmann interpretations (chapters 1 and 5). Additionally, we have seen how the idea of entropy can effectively be used in evolution or social theories.

The most compelling lesson from this law is the rejection of determinism and eternity offered in classical era science and earlier by religious teachings. Entropy teaches us that the fundamental

Fig. 6.4. *The idea of eternity is appealing to humans, but the second law of thermodynamics illustrates it is an illusion. Image courtesy of Robbert van der Steeg via Wikimedia Commons.*

laws of nature can only be expressed as probable states, not with absolute certainty, and coupled with the second law of thermodynamics, it tells us that eternity is an illusion (Fig. 6.4). This scenario may not be very comforting to many of us, but as Carl Sagan noted in his popular book, Cosmos (1985), "The universe is not required to be in perfect harmony with human ambition."

Bibliography

Bayliss, J. T. (1970). *The second law of thermodynamics: An informal review*. Richmond, VA: Virginia Commonwealth University.

Bohm, D., & Peat, F. D. (2000). *Science, order, and creativity*. London: Routledge.

Boltzmann, L. (1974). The second law of thermodynamics. In McGuinness, B. (ed.) *Theoretical physics and philosophical problems* (pp. 13–32). New York: Springer-Verlag LLC.

Brahma, I. (2017, June). Board #81: Rethinking the macroscopic presentation of the second law of thermodynamics, Paper presented at 2017 ASEE Annual Conference & Exposition, Columbus, Ohio. Retrieved from https://peer.asee.org/27933

Clausius, R. (1870). On a mechanical theorem applicable to heat. *The London, Edinburgh, and Dublin Philosophical Magazine and Journal of Science 440*(265), 122–127.

Did Obama Really Mean 'Social Darwinism'? (2012, April 5). *CNBC*. Retrieved from https://www.cnbc.com/id/46964102

Eddington, A. S. (1933). *The expanding universe*. Cambridge: Cambridge University Press.

Eddington, A. S. (1928). *The nature of the physical world*. New York: Macmillan Co.

Eddington, A. S. (1935). *New pathways in science*. Cambridge: Cambridge University Press.

Greven, A., Keller, G., & Warnecke, G. (2003). *Entropy*. Princeton: Princeton University Press.

Hayflick, L. (2007). Biological aging is no longer an unsolved problem. *Annals of the New York Academy of Sciences, 1100*(1), 1–13. doi:10.1196/annals.1395.001

Hayflick, L. (2007). Entropy explains aging, genetic determinism explains longevity, and undefined terminology explains misunderstanding both. *PLoS Genetics, 3*(12): e220.

Hockey, T., et al. (eds.). (2014). *Biographical encyclopedia of astronomers*. New York, NY: Springer.

Kostic, M. (2016). Entropy generation results of convenience but without purposeful analysis and due comprehension—Guidelines for authors. *Entropy 18*(1), 28. Retrieved from https://www.mdpi.com/1099-4300/18/1/28; https://doi.org/10.3390/e18010028

Kovalev, A. (2015). Misuse of thermodynamic entropy in economics. *Energy 100* (2016 April), 129–136. Retrieved from http://ssrn.com/abstract=2635663

Lambert, F. L. (2007). Entropy and the second law of thermodynamics. Retrieved from http://www.entropysite.com.

Lauwerys, J. A. (1952). Herbert Spencer and the scientific movement. In Judges, A. V. (ed.). *Pioneers of English Education*. London: Faber.

Monod, J. (1972). *Chance and necessity: An essay on the natural philosophy of modern biology*. New York: Vintage Books.

Pinker, S. (2017). 2017: What scientific term or concept ought to be more widely known? *Edge*. Retrieved from https://www.edge.org/response-detail/27023

Sagan, C. (1985). *Cosmos*. New York, NY: Ballantine Books.

Schrödinger, E. (1955). *What is life? The physical aspect of the living cell.* Cambridge: Cambridge University Press.

Thompson, D. W. (1942). *On growth and form* (2nd ed.). Cambridge: Cambridge University Press.

Young, A. P. (1948). *Lord Kelvin*. London: Longmans.

CHAPTER 7

The equation that unraveled the dual nature of light

Science cannot solve the ultimate mystery of nature. And
that is because, in the last analysis, we ourselves are a part of
the mystery that we are trying to solve.
— *Max Planck, Where is science going? (1932)*

Abstract

*At the end of the nineteenth century, many people thought there was
nothing new to be discovered in physics. All that remained was more
and more precise measurement. In 1900, Max Planck postulated
that electromagnetic energy was emitted not continuously but by
discrete portions or quanta. Quantum mechanics was born! Planck
calculated the energy E of each photon by multiplying Planck's
constant h with the frequency of the wave. The equation not only*

calculated the energy of particles and waves but also emphasized the fact that matter and energy were quantized, and consequently, space and time were also quantized.

At the end of the nineteenth century, physics appeared to have hit the wall, reasoning that there was nothing new to be discovered in physics and all that remained was more and more precise measurement. This statement has generally been attributed to William Thomson (aka Lord Kelvin; 1824–1907). Paradoxically, not only was the reasoning proven wrong but there was also something big waiting to happen in physics. As the twentieth century began, Max Planck entered the arena of physics with a new and radical explanation for an existing phenomenon, which eventually paved the way for quantum mechanics. It is not an exaggeration to say that this new idea shattered the foundations of classical physics, which had been built on determinism. The core issue that Planck successfully addressed, and that classical physics had failed to address, was the explanation for observed radiation from a blackbody.

A perfect blackbody is an ideal object that absorbs all the light it receives and does not reflect any of it. The name blackbody comes from the fact that such an object appears to be perfectly black since it does not reflect any light. However, when a blackbody is heated to a high temperature, it will begin to emit radiation. According to classical physics, an object in thermal equilibrium emits radiation of a given intensity at all possible wavelengths and emits more when frequency increases. As the frequency of the emitted light increases near the ultraviolet end of the spectrum, the classical explanation is that the blackbody should radiate an infinite amount of energy at high frequency. As you can imagine, this would violate the law of conservation of energy. Since the blackbody received only finite energy, it cannot emit an infinite amount of energy.

Max Planck offered a solution to this problem, and it turned out to be one of the fundamental building blocks of quantum mechanics. Planck asserted that the energy was emitted in discrete packets, or quanta, rather than in continuous ranges as predicted by classical physics. Figure 7.1 shows the intensity of emitted radiation as a function of wavelength. The spectrum of blackbody radiation shows that, at some wavelengths, energy is higher than at others. Since wavelength and frequency are inversely proportional, as the frequency increases so does the energy. Since ultraviolet light has a higher frequency, it should have more energy (ultraviolet has higher frequency than visible light). By calculating the total amount of radiated energy, it can be shown that a blackbody would release an infinite amount of energy in the ultraviolet range. In other words, the maximum wavelength emitted by a

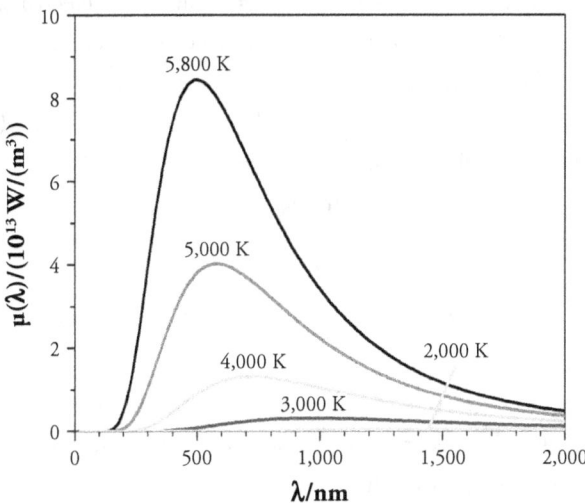

Fig. 7.1. *The intensity of radiation is a measure of the energy emitted per unit area. A plot of the intensity of blackbody radiation as a function of wavelength for an object at various temperatures. Image courtesy of Wikimedia Commons.*

blackbody radiator is infinite, and the classical physics explanation couldn't match the observed shape of the spectrum.

From the blackbody radiation spectrum, we can infer some of these interesting facts:

- The blackbody spectrum depends only on the temperature of the object. This means the composition of the blackbody is irrelevant. Different objects at the same temperature should show the same spectrum.
- The higher the temperature, the higher the energy emission.
- The peak wavelength of the blackbody spectrum becomes shorter as the temperature increases.

In order to explain the blackbody radiation phenomenon, Planck rejected the classical idea that each frequency of vibration should have the same energy and assumed that electrons that vibrated with different frequencies did not share energy equally. Planck postulated that the energy came in small packets, each of which he called a quantum, and the amount of energy (E) of a vibrating particle depended on the frequency (f) of vibration and could be calculated using the equation:

$$E = hf$$

where h is called Planck's constant. Its value is about 6.626 × 10^{-34} Js.

So, how did this idea become the foundation of quantum mechanics? Planck explained that an electron vibrating with a frequency f could only have energy as an integer multiple of hf, that is, 1 hf, 2 hf, 3 hf, 4 hf, etc. Additionally, an electron should have a minimum of one quantum of energy for it to vibrate and emit energy. Otherwise, it would not vibrate at all and could not produce any light.

So, it turns out that atoms can only take certain specific values while absorbing or emitting energies. In other words, atomic vibrations are *quantized*. This idea, when generalized, led to some radical shifts in the way physicists were thinking about energy, and on a grand scale, it changed the way we viewed our universe. (In fact, all phenomena in submicroscopic systems exhibit quantization, including our notion of space-time at the fundamental level.) So, with his new idea of quantization, Planck opened the door to a completely new world in physics, namely, quantum mechanics. The 1918 Nobel Prize in physics was awarded to Max Karl Ernst Ludwig Planck "in recognition of the services he rendered to the advancement of physics by his discovery of energy quanta." Needless to say, this was a turning point in the history of physics. The idea of quantization had immense ramifications in a variety of fields, from optics to astronomy.

At this point, we need to look at the relationship between temperature and frequency or wavelength. Since wavelength and frequency are inversely proportional, the radiation emitted by an object would be towards the low end of the frequency (longer wavelength) and vice versa. You can see this relationship on the above spectrum (Fig. 7.1), and it had some huge practical benefits. This relationship between temperature and wavelength would allow astronomers to know the temperature of stars or other objects which are quite distant from us and cannot be measured directly.

The temperature (T, in K) of an object and the wavelength at which it emits the most radiation (λ_{max}, in m) are related: $\lambda_{max} T = 2.898 \times 10^{-3}$ m K. This relationship is known as Wien's displacement law (named after the German physicist Wilhelm Wien who discovered it). In astronomy, this relationship is very useful as it enables astronomers to calculate the temperature of stars and other objects by simply measuring the emitted wavelength at which their emission is most intense. In fact, it is the vibration of the electrons that produces light or any other electromagnetic radiation. Apparently, when an object is hot, the electrons vibrate more.

The following chart (Fig. 7.2) helps us to categorize different stars by their luminosity and temperature. Notice that as the temperature becomes higher, the source emitting the radiation moves towards blue. In other words, blue stars are hotter than red ones.

One can only imagine how important the relationship between color and temperature is for astronomers when it comes to cataloging astronomical objects. Although all astronomical objects are only approximate blackbodies, astronomers have discovered an almost perfect blackbody in what they call the cosmic microwave background (CMB). This is the leftover radiation from the Big Bang that happened about 14 billion years ago.

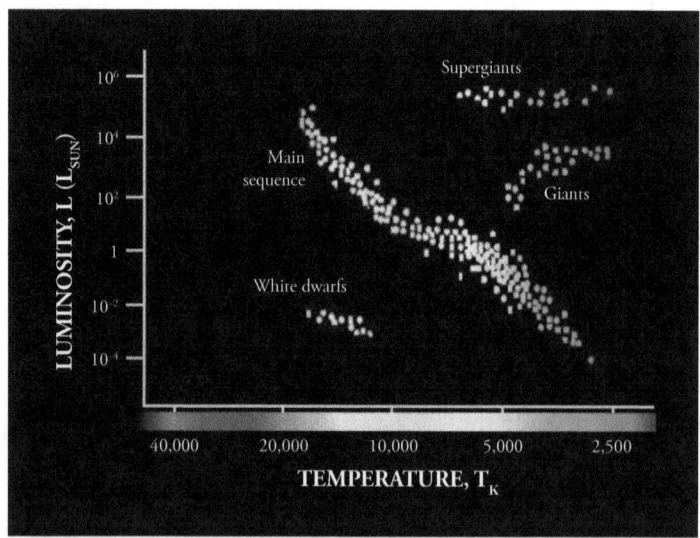

Fig. 7.2. *The relation between temperature and color (wavelength) of stars is obvious on this chart. Image courtesy of Wikimedia Commons.*

(We discussed the Hubble constant and how it helped to calculate the age of the universe in chapter 4.) It has been estimated that CMB originated in the early universe, about 380,000 years after the Big Bang, and pervades the entire universe even today. As the CMB has cooled over the years, scientists estimate the current temperature of CMB to be around 2.7 Kelvin, which is just above absolute zero (Fig. 7.3).

Planck's idea and equation were originally meant to explain the blackbody radiation emitted by an ordinary object that acted as a blackbody in laboratory setup. Nevertheless, this idea became one of the most revolutionary laws in modern physics. In fact, Albert Einstein used the same principle later on to explain the

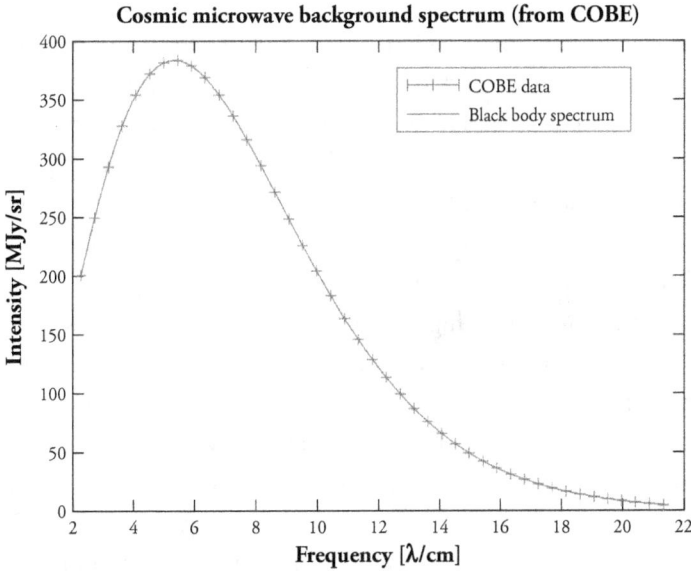

Fig. 7.3. *The most perfect blackbody spectrum ever measured. The spectrum of the CMB measured by the COBE satellite. The datapoints are indistinguishable from theory, and the error bars of the measurement are too small to plot. Image courtesy of Wikimedia Commons.*

photoelectric effect for which he was awarded the Nobel Prize in physics in 1921. Since the advent of the quantization idea, a new era in physics has begun, and quantum mechanics is considered to be one of the most successful theories the world has seen.

It is interesting to note here that once the determinism advocated by classical physics had waned, certainty and absolute truth that had ruled the world for centuries became unfavorable artifacts. We have learned to live with uncertainty and probability, and the implications of this idea have even altered our notions about the nature of light.

The dual nature of light

Since ancient times, philosophers and thinkers have pondered the fundamental units that make up our world. The ancient Greek philosopher Democritus held the view that light was composed of particles like everything else in the universe. Later on, Isaac Newton held a similar view, called the corpuscular theory, which described light as tiny particles known as corpuscles. These corpuscles, on emission from a source of light, travel in a straight line with high velocity. However, during this time, Christian Huygens, a Dutch physicist, proposed his own principle in which he described a geometrical way of understanding the behavior of light waves—in essence, the wave theory of light.

Although both of these views on light were entertained, Newton's corpuscular view on light dominated the world of physics for a considerable amount of time. Earlier, we discussed Planck's idea of light as a quantum to resolve the blackbody radiation and how that propelled a new way of thinking in physics—quantum mechanics. Remember, a quantum of light is called a photon, and Einstein used this particle nature of light for his explanation of the photoelectric effect.

Experiments in physics, and several phenomena exhibited by light, established the fact that light could behave simultaneously as a particle and as a wave. Now the question became: "If light is a wave and a particle depending on the observation, what is the true nature of matter?" One of the most famous experiments in physics, called the double slit experiment, demonstrated that some particles of matter, like electrons, behave like waves.

As you can imagine, the dual nature of matter and light was so radical in physics that its repercussions are felt even today. The wave-particle duality represents the mysterious ways by which nature presents itself to us. It sounds crazy to accept the strange fact that *everything is a wave and a particle!* In our journey from classical physics to quantum mechanics, we had no choice other than to hold this rather embarrassing fact.

Although a variety of experiments has successfully observed both the particle- and wave-like behaviors of light, the experiments have never been able to observe both at the same time. Since the days of Einstein, scientists have been trying to directly observe both of these aspects of light at the same time. Now, scientists at the Swiss Federal Institute of Technology, in Lausanne, Switzerland (EPFL) have succeeded in capturing the first-ever snapshot of this dual behavior (Piazza et al., 2015). Taking a radically different experimental approach, EPFL scientists have now been able to take the first ever snapshot of light behaving both as a wave and as a particle, as shown below (Fig. 7.4).

If reality is ambiguous, as expressed by the dual nature of light, why should there be a general perception to know everything in a definite sense. Science, probably, does not have an absolute solution or a final answer for many of the observed

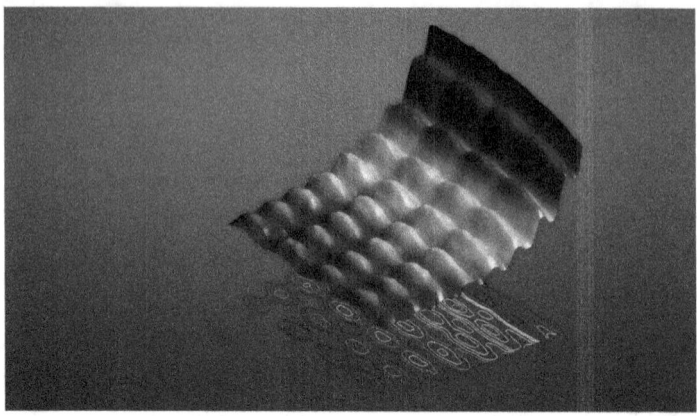

Fig. 7.4. *Light as both wave and particle. Image courtesy of École Polytechnique Fédérale de Lausanne (EPFL), Switzerland and Dr. Fabrizio Carbone, Laboratory for ultrafast microscopy and electron scattering (LUMES), EPFL: https://actu.epfl.ch/news/the-first-ever-photograph-of-light-as-both-a-parti/.*

phenomena. In fact, that contributes to the strength of science as it is an ongoing process. Science is a verb and not a noun as many would like to think. In his book, *The Demon-Haunted World: Science as a Candle in the Dark*, astronomer Carl Sagan noted that:

> ...even laws of Nature are not absolutely certain. There may be new circumstances never before examined—inside black holes, say, or within the electron, or close to the speed of light—where even our vaunted laws of Nature break down and, however valid they may be in ordinary circumstances, need correction.

However, for now, there is no substitute for Planck's equation, and to see how it affects our contemporary world, let us look at Planck's constant again.

Planck's constant

The legacies of Max Planck remain large in physics, with many of the physical quantities bearing his name, for example, Planck length, Planck time, etc., in addition to Planck's constant (h) that we discussed briefly at the beginning of this chapter.

Planck's constant has had profound ramifications in three important areas: our technology, our understanding of reality, and our understanding of life itself (Stein, 2011). The other constants that match such criteria and are universal are the gravitational constant (G) and the speed of light (c). It's a well-known fact that knowledge of the Planck's constant is crucial in electronic system designs such as integrated circuits. But, recently, it has also helped physicists to set a new standard for the kilogram, the basic unit of mass.

Around 1960, the International System of Units (SI) was established, and countries around the world accepted metrology institutions to continually refine the official definitions of our seven base units of measure: the meter (length), kilogram (mass), second (time), ampere (electric current), kelvin (temperature), mole (amount of substance), and candela (luminosity). We can derive other units from these base units so they are considered fundamental units. For example, velocity is m/s which shows it is mathematically equal to distance covered in a direction divided by time.

According to the SI standard, the kilogram equals 1,000 grams as defined by the international prototype kilogram (IPK) of platinum-iridium kept at the International Bureau of Weights and Measures in Sèvres, France (Fig. 7.5). In 2018, the General Conference on Weights and Measures voted to stop the IPK from being used as the definition of the kilogram (Fig. 7.6).

Fig. 7.5. *The international prototype kilogram is a cylinder of platinum and platinum-iridium alloy, which is kept at the International Bureau of Weights and Measures (BIPM) near Paris. Image courtesy of BIPM.*

Instead, the Planck's constant will be used to define the kilogram, and this will be effective from May 20, 2019.

To understand how the Planck's constant can be used to define the kilogram, look at the unit of this constant which is joule-second, or J·s. Here joule (J) is the unit of energy and second (s) is the unit of time. In fact, a joule is equal to a kilogram multiplied by meters squared divided by seconds squared (kg·m²/s²).

This relationship shows that the Planck constant can be connected to the kilogram with some calculation, and that is exactly what the new definition of kilogram is planning to achieve—linking the Planck constant to the kilogram. Defining a term such as kilogram in association with a fundamental constant like

Fig. 7.6. *The NIST-4 watt balance has measured Planck's constant to within 34 parts per billion, demonstrating that the high-tech scale is accurate enough to assist with 2018's planned redefinition of the kilogram. Image courtesy of J. L. Lee/NIST.*

the Planck's constant, rather than in association with manmade objects like IPK, would ensure the utmost precision that we can provide in maintaining a basic unit like the kilogram. The unit of distance, which is the meter, has already been defined using another physical constant (speed of light) as follows: the length of the path travelled by light in a vacuum during a time interval of 1/299,792,458 of a second. As mentioned earlier, physicists consider Planck's constant as a universal constant similar to the speed of light. The universal constants have a deeper meaning than the numerical values they represent. Any sentient beings quantifying their world in any part of the universe must figure out the same value for these constants; hence they are universal.

Now, let us look at the other Planck scales referred to as Planck length and Planck time. Although these quantities do not have any significance in current scientific endeavors due to their extremely small sizes, they teach us about the limits of human knowledge when dealing with the microscopic world. It has been assumed that the known laws of physics break down when we reach the boundaries set by the Planck scales.

Let's consider the unit of Planck length to see how interesting these scales are and why they set some limits on our understanding. The Planck length, the smallest measurement of length, is theoretically calculated to be 1.6×10^{-35} meters. To get a better perception, a proton is about 100 million trillion times larger than the Planck length.

Similarly, Planck time is estimated to be 5.4×10^{-44} seconds. This is the time it would take a photon travelling at the speed of light to cross a distance equal to the Planck length. Also known as "quantum of time," it is the smallest measurement of time, and any smaller division of time has no meaning. In a sense, before Planck time, all of the four fundamental forces were presumed to have been unified into one force. All matter, energy, space, and time originated from this primordial state, the beginning of our universe.

Does the Planck scale set the limit for the universe? In a way, it tells us space is not an infinitely divisible continuum as suggested by the special theory of relativity. Theoretically, Planck length gives the size of the smallest possible piece space can be. In other words, at distances of 10^{-35} meters and times of 10^{-44} seconds, none of our known theories would be able to explain the science any more. Thus, the Planck scale sets the universe's minimum limit, beyond which the laws of physics break down.

Only speculations exist regarding the Planck scales as no real measurable science can be done at this level. Our current understanding is that the universe is governed by four fundamental forces, namely, gravity, the weak force, electromagnetism, and the strong force. Some theoretical physicists believe all these fundamental forces might merge into one force at this scale. The Planck equation opened up a completely new world in physics, and it also seems to have set a limit on human understanding of the material universe.

We can definitely come to the conclusion that Max Planck revitalized humanity's urge to explore the limits. The contribution of Planck, in Einstein's own words, "...became the basis of all twentieth-century research in physics and has almost entirely conditioned its development ever since. Without this discovery, it would not have been possible to establish a workable theory of molecules and atoms and the energy processes that govern their transformations. Moreover, it has shattered the whole framework of classical mechanics and electrodynamics and set science a fresh task: that of finding a new conceptual basis for all physics." There is no doubt that the Planck constant and scales continue to guide us to the unknown realms of our universe.

Planck died in 1947. Decades after his demise, in his honor, the scientific community named a satellite after him that explored the mysteries of the origin of the universe.

Planck satellite

Planck was the European Space Agency's (ESA) space-based observatory and was in full operation from 2009 to 2013. The main goal of its mission was to study the cosmic microwave background—the relics of radiation left over from the Big Bang—across the whole sky at greater sensitivity and resolution than ever before. According to the ESA, Planck resembled a time machine, giving astronomers insight into the evolution of the universe since its birth nearly 14 billion years ago.

We learned in chapter 4 how Hubble's findings led to the Big Bang theory, which is considered as the cornerstone of modern cosmology. One of the most convincing pieces of evidence for the Big Bang theory came from the cosmic microwave background (CMB) radiation predicted by the Big Bang model. The CMB is the oldest light or radiation in the universe that emerged about 380,000 years after the bang but has cooled since then. The nearly uniform low-temperature glow that fills the universe remains in the microwave frequency range of the electromagnetic spectrum (Fig. 7.7). These ancient photons, which emerged when the universe was much younger, are present even in our own living rooms. A small part of the noise you detect from your television or radio is from the CMB.

It should be mentioned here that in the 1940's physicist George Gamow postulated the existence of the cosmic microwave background on theoretical grounds, but CMB was first detected accidentally in 1964 by Arno Penzias and Robert Wilson, for

Fig. 7.7. *The details in the Big Bang's leftover glow have been progressively better and better revealed by improved satellite imagery. The latest, final results from Planck provide us with our most informed picture of the universe of all time. Image courtesy of NASA/ESA and the COBE, WMAP, and Planck teams.*

which they were awarded the Nobel Prize in physics in 1978. Many attempts have been made to map CMB in recent times.

The first space-based, full-sky map came from NASA's Cosmic Background Explorer (COBE) mission, which launched in 1989 and ceased science operations in 1993. It confirmed the predictions of the Big Bang model and gave us a baby picture of the universe. The Wilkinson Microwave Anisotropy Probe (WMAP) launched in June 2001 and gave more detailed pictures of CMB. In chapter 4, we discussed in detail the Hubble constant and its significance in astrophysics. According to the Planck satellite's observations of the distant early universe, the Hubble constant is 67 kilometers per second for every 1 million parsecs of separation in space (67 km/s/Mpc). (One parsec is about 3.26 light-years.)

Figure 7.7 shows the CMB map provided by the COBE, WMAP, and Planck satellites. Planck's map greatly improved our understanding of the universe. In a universe where there are more galaxies than there are humans on this planet, there is no lack of wonders. The eons may come and go, but the marvels of our existence will remain—and so does the beauty of the Planck equation.

Bibliography

Einstein, A. (2005). *Albert Einstein: Out of my later years: the scientist, philosopher and man portrayed through his own words.* Edison, NJ: Castle Books.

Feynman, R. (1990). *The character of physical law.* Cambridge, MA: MIT Press.

Hawking, S. W., & Penrose, R. (1996). *The nature of space and time.* Princeton, NJ: Princeton University Press.

Hinshaw, G., Larson, D., Komatsu, E., Spergel, D. N., Bennett, C. L., Dunkley, J. ... Wright, E. L. (2013, October 1). Nine-year Wilkinson microwave anisotropy probe (WMAP) observations: Cosmological parameter results. *The Astrophysical Journal Supplement*, *208*(2), 25 pp. Retrieved from http://adsabs.harvard.edu/abs/2013ApJS..208...19H

Milonni, P. W. (1984). *Wave-Particle duality of light: A current perspective.* Dordrecht: Springer.

Papageorgiou, N. (2015). *The first ever photograph of light as both a particle and wave.* École Polytechnique fédérale de Lausanne (EPFL). Retrieved from https://actu.epfl.ch/news/the-first-ever-photograph-of-light-as-both-a-parti/

Peacock, J. A. (1998). *Cosmological physics.* Cambridge: Cambridge University Press.

Peebles, P. J. E., Page, Jr., L. A., & Partridge, R. B. (eds.). (2009). *Finding the Big Bang.* Cambridge: Cambridge University Press.

Piazza, L. et al. (2015). Simultaneous observation of the quantization and the interference pattern of a plasmonic near-field. *Nature Communications 6*, 6407. doi: 10.1038/ncomms7407.

Planck. (n.d.). NASA/IPAC Infrared Science Archive. Retrieved from https://irsa.ipac.caltech.edu/Missions/planck.html

Planck, M. (1932). *Where is science going?* New York: W.W. Norton.

Planck Collaboration (Ade, P. et al.). (2014). Planck 2013 results. XIV. Zodiacal emission. *Astron. Astrophys. 571*, A14, arXiv:astroph/1303.5074.

Planck Collaboration (Arnaud, M. et al.). (2014, March 20). Planck 2013 results. XVI. Cosmological parameters. Retrieved from https://arxiv.org/abs/1303.5076

Preuss, P. (2014, May 23). Planck mission updates: The age of the universe and what it contains. Berkeley Lab. Retrieved from https://newscenter.lbl.gov/2013/03/21/planck-results/

Sagan, C. (1995). *The demon-haunted world: Science as a candle in the dark*. New York: Random House.

Sharlin, H. I. (1979). *Lord Kelvin: The dynamic Victorian*. University Park: Pennsylvania State University Press.

Shivni, R. (2016). The Planck scale. *Symmetry*. Retrieved from https://www.symmetrymagazine.org/article/the-planck-scale

Siegel, E. (2018). How the Planck satellite forever changed our view of the universe. *Forbes*. Retrieved from https://www.forbes.com/sites/startswithabang/2018/07/19/how-the-planck-satellite-changed-our-view-of-the-universe/#4c5e7c9a7ad2

Stein, J. (2011). Planck's constant: The number that rules technology, reality, and life. *PBS NOVA*. Retrieved from https://www.pbs.org/wgbh/nova/article/plancks-constant/

CHAPTER 8

The equation that reveals the interplay between mass and energy

It followed from the special theory of relativity that mass
and energy are both but different manifestations
of the same thing—a somewhat unfamiliar conception
for the average mind.
—*Albert Einstein, Relativity: The Special and the General
Theory (1952)*

Abstract

*Einstein's mass energy equivalence relation is the most famous equa-
tion in the history of physics. Einstein himself described the equivalence of mass and energy as "the most important upshot of the special
theory of relativity." This equation provides a clean explanation of
the relation between mass and energy. It also explains what powers*

the stars, including our own Sun. Yes, it provided a framework for the construction of the atomic bomb. *The particle accelerators are another arena where we can see this equation at play.*

The equation, $E = mc^2$, without any doubt, is the world's most famous equation. It is one of the most recognizable equations and has transcended cultural and geographical barriers. Everyone easily recognizes the symbols (m – mass, c – speed of light) in this simple mathematical equation, yet it is more complex than anyone would think. This deceivingly simple equation is the essential outcome of the special theory of relativity which, along with the general theory of relativity, shot Einstein to fame.

In 1905, Albert Einstein published his paper "Does the inertia of a body depend upon its energy content?" in the journal *Annalen der Physik*. It was the last of four papers he submitted that year to this journal. The paper presented the relationship between energy and mass that would eventually lead to the mass-energy equivalence formula, $E = mc^2$. It should be noted here that a few scientists before Einstein had proposed the mass-energy relationship in various contexts, but it was Einstein who elevated the mass-energy equivalence to the level of a fundamental principle in modern physics. This relationship followed from the relativistic symmetries of space and time that he introduced as part of the special theory of relativity, yet the implications of this equation were much deeper than anyone had previously thought. To understand the origin of this equation, we need to briefly mention the special theory of relativity here.

The two postulates of the special theory of relativity are:

- The laws of physics take the same form in all frames of reference, moving with constant velocity with respect to one another.

- The speed of light in a vacuum is the same for all observers, regardless of the motion of the light source.

It needs to be mentioned here that special relativity describes objects that move in an inertial frame of reference. An inertial frame of reference has a constant velocity, which means an object is either moving at a constant speed in a straight line or it is standing still (velocity zero). In other words, it is a non-acceleration frame of reference in which Newton's laws of motion hold. But, where is the mass-energy equivalence here? As Einstein himself put it: "It followed from the special theory of relativity that mass and energy are both but different manifestations of the same thing—a somewhat unfamiliar conception for the average mind." Remember Einstein was a theoretical physicist, so doing experiments and reaching conclusions was not his style. He was essentially a paper-pencil scientist. As such, other scientists had derived this equation mathematically using some of the existing observations, and he knew about some of the previous attempts in this regard by other scientists. The mathematical derivation of this equation is beyond the scope of this book, but look at Figure 8.1 in which Einstein himself is presenting the derivations in the special theory of relativity for an audience.

As mentioned before, the year 1905 was a miracle year for Einstein. Later in his life, Einstein described how, at the age of 16, he imagined he was chasing after a beam of light and how that thought experiment had played a memorable role in his development of special relativity. He called this thought experiment the "germ of the special relativity theory." He imagined a scenario in which he was pursuing a beam of light in a vacuum, going as fast as light. In this case, Einstein thought, that light should appear stationary or frozen, since both he and the light were going at the same speed.

Fig. 8.1. *Einstein deriving special relativity, for an audience, in 1934. Image is in the public domain.*

Einstein expanded on this in his 1949 autobiographical notes:

> If I pursue a beam of light with the velocity c (velocity of light in a vacuum), I should observe such a beam of light as an electromagnetic field at rest though spatially oscillating. There seems to be no such thing, however, neither on the basis of experience nor according to Maxwell's equations. From the very beginning it appeared to me intuitively clear that, judged from the standpoint of such an observer, everything would have to happen according to the same laws as for an observer who, relative to the Earth, was at rest. For how should the first observer know or be able to determine, that he is in a state of fast uniform motion? One sees in this paradox the germ of the special relativity theory is already contained.

The conflict between what he thought to be true and the existing scientific wisdom of that time played a huge role in shaping his thinking and eventually led to the special theory of relativity. The beam of light carries a certain amount of momentum, and the velocity of the object itself is unchanged. Since momentum is conserved (a well-known law in physics), this is only possible if the object's inertial mass has changed, and indeed, Einstein found that the change in inertial mass is equal to $\Delta L/V^2$. Although we are most familiar with $E = mc^2$ (Fig. 8.2), Einstein didn't quite say it that way in the original paper. It was actually in the form of a sentence in German: "If a body gives off the energy L in the form of radiation, its mass diminishes by L/V." Notice that, in the original paper, Einstein uses V instead of c for the speed of light, and L instead of E for energy. In addition, if we rewrite this relation using E and c instead of L and V, it should lead to the current form of the equation. Einstein concluded that an object's inertial mass is proportional to its energy content with c^2 added as a proportionality constant.

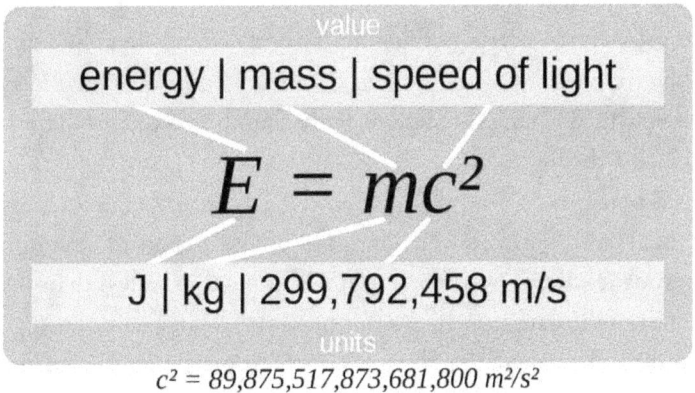

$$c^2 = 89{,}875{,}517{,}873{,}681{,}800 \ m^2/s^2$$

Fig. 8.2. *The equation* $E = mc^2$ *on display with units and constants. Image courtesy of Wikimedia Commons.*

The interplay between mass and energy was beyond any experimental verification in the early twentieth century. Today, we know from the actions that take place in particle accelerators (where fundamental particles are collided at high speed to generate other exotic particles), and from phenomena like nuclear fission or fusion, how significant a role this mass-energy relation plays.

Converting mass into energy is the goal of scientists pursuing nuclear fission or fusion (more on that in the section "Star in a jar"). In simplest terms, nuclear fission is the splitting of an atomic bond that keeps the nucleons together in an atom. A huge amount of energy is needed to accomplish the separation of two nucleons; however, it can produce much more energy in the process. We learn from history that this tremendous task was undertaken during the Second World War and culminated in the making of the atomic bomb, code-named Manhattan Project.

It should be mentioned here that, although Einstein himself had written to the U.S. government asking it to fund research into atomic energy during the Second World War, his own involvement was quite limited. Einstein's equation didn't have much to do with the design of the bomb except that it provided a theoretical framework in which mass-energy conversion was possible.

As you can see from Figure 8.3, the fission of uranium atoms results in two new nuclei with a total mass less than the mass of the parent nucleus. This missing mass appears as energy and can be calculated using $E = mc^2$. In principle, this energy can be harnessed for useful (nuclear reactors) or destructive (atomic bomb) purposes. Even in our daily lives, a deep link exists between this equation and the way we consume energy.

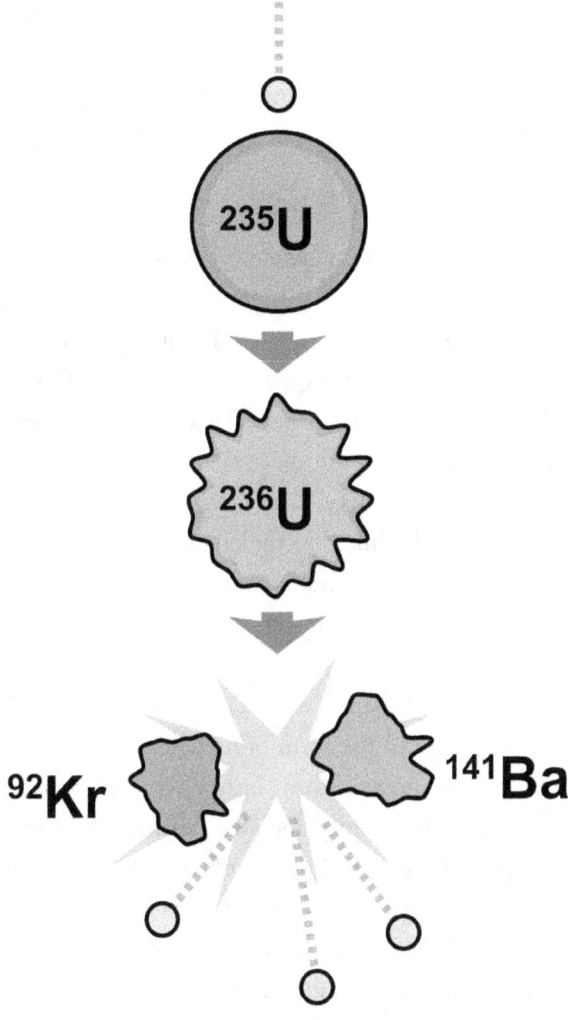

Fig. 8.3. *Simple diagram of nuclear fission. In the first frame, a neutron is about to be captured by the nucleus of a U-235 atom. In the second frame, the neutron has been absorbed and briefly turned the nucleus into a highly excited U-236 atom. In the third frame, the U-236 atom has divided into fission fragments (Ba-141 and Kr-92) and three neutrons, all with very large amounts of kinetic energy. Image courtesy of Wikimedia Commons.*

The fossil fuels that we mostly depend on come from prehistoric organic material buried under the Earth. But they received the energy from the Sun before, and the Sun's energy comes from nuclear fusion, which also can be computed using the very same equation, $E = mc^2$.

In addition to its practical purposes, this equation also provides a context in which to generate a philosophical view of the universe where everything is interconnected, and the material world is a manifestation of the same underlying principle. In the last chapter, we discussed the particle-wave duality and how these visibly different aspects (particles and waves) are essentially exhibited by everything in the universe. Similarly, mass and energy are manifestations of the same principle. Our material world is very closely linked to the energy that the universe creates. In physicist Ludwig Boltzmann's words, "Available energy is the main object at stake in the struggle for existence and the evolution of the world."

In the world of subatomic processes, the mass of particles can change into energy in the form of light, heat, or motion. Likewise, energy can also transform into mass. Particle accelerators exploit this idea by smashing together fast-moving particles. The high energy of these collisions transforms them into new particles, which can have a much greater mass than the ones that originally collided. We all know about Higgs Bosons (discovered in 2012) and the famous Large Hadron Collider (LHC) operated by CERN, the European Organization for Nuclear Research. But, perhaps, we have not heard of the particle accelerator that was never built and that, in fact, would have been much larger (more energetic) than the LHC if it had been completed on U.S. soil.

Too big to succeed—the abandoned Desertron is a bleeding scar on the heart of American particle physics

In the 1980s, even before construction of the now-famous Large Hadron Collider was approved by CERN (European Organization for Nuclear Research), American physicists began an ambitious plan to build a particle collider, to be specific, the largest particle accelerator ever planned, outside Waxahachie in Texas. Nevertheless, in 1993, citing rising costs, the U.S. government decided to cancel the partially built Superconducting Super Collider (SSC) facility that would eventually leave its European counterpart as the sole candidate for high-energy particle physics experiments. The story of this abandoned accelerator is not just the case of a neglected project that was too big to succeed, but in a deeper sense, it epitomizes the symptoms plaguing the state of American particle physics lately.

If it had been completed, the Super Conducting Super Collider, also known as Desertron, with a ring circumference of 54 miles, would have easily dwarfed the LHC that has a ring circumference three times less than SSC, both in terms of size and energy. Particle accelerators do not come cheap, and the bigger and more powerful the machine, the more exotic particles will be unveiled to us. Several physicists, and for valid reasons, believe that the discovery of the Higgs Boson (the God particle) would have happened on U.S. soil well before it occurred at LHC had Desertron been up and running as planned.

However, after spending two billion dollars and completing 14 miles of tunnels, the massive SSC facility, which could have been a beacon of American scientific eminence, was abandoned under the dusty Texas soil. The decision to kill the SSC project was a horrifying experience, not only for many veteran scientists but also for a new generation of particle physicists. For them, the partially built tunnels were reminiscent of a long bleeding scar on the heart of American particle physics.

From the detection of antiprotons at the Lawrence Berkeley National Laboratory's Bevatron in California in the mid-50s to the discovery of the top quark in 1995 at Fermilab's Tevatron (which is roughly 1,000 times more powerful than the Bevatron), all were testaments to America's once glorious and highly evolving particle physics adventures. That momentum, it seemed, ceased with the cancellation of the SSC project.

During the congressional debate about the future of the partially constructed SSC, many physicists, including Nobel-Prize-winning physicist Leon Lederman who coined the term *the God particle*, were involved in a campaign to persuade the U.S. government to continue funding for the Superconducting Super Collider. Lederman sought to promote awareness and support of this project through the popular science book, *The God Particle: If the Universe Is the Answer, What Is the Question?*, which he wrote along with Dick Teresi. Nonetheless, the U.S. Congress voted to defund the massive science project, leaving the SSC facility abandoned and neglected for decades (Fig. 8.4).

Apparently, this marked the long and steady decline of American particle physics that continues, to date, with significant budget cuts to the high energy physics program. The federal support for particle physics has declined in real terms by more than 50% during the last two decades. In 2012, the partially constructed SSC facility was purchased by Magnablend, a local chemical manufacturer, for its operations.

Particle accelerators are not just machines that display the mass-energy interplay, but they also allow us to peer more deeply

Fig. 8.4. *Tunnels left behind in Texas as the SSC project was canceled. U.S. government photo is in the public domain.*

into the laws of nature that govern everything. In that sense, the SSC was a machine from the future which, had it been completed, would have remained the most powerful accelerator in the world, revealing many unknowns that remain unknown to us even today.

Perhaps the SSC was too big to succeed, and any future prospect of having such a big machine in the American particle physics landscape is bleak. Yet, it was a lost opportunity and has already dealt a blow to the training and inspiring of a new generation of particle physicists in the United States. Strangely enough, in 2015,

Lederman auctioned off his 1988 Nobel gold medal to meet his costly medical bills following a dementia diagnosis.

Some physicists speculate that the United States no longer considers particle physics as important as it used to. It seems that America has exited the particle physics highway it traveled at high speeds in the past. Or could we say the curse of the abandoned Desertron continues? Whatever our perception, there is no wizard available, at least in the near future, to cast a spell on American particle physics.

Although the relation between mass and energy is obvious in nature and laboratories, energy production from mass is not an easy task. The currently operating nuclear plants are based on the fission process, and the public fear of radiation fallout has somewhat diminished their role in energy production in many parts of the world. However, nuclear fusion seems to have a huge potential in solving world energy problems, but the technical difficulty in achieving nuclear fusion is still a big barrier to providing abundant clean energy to the world. But that won't stop scientists and engineers from trying to accomplish this dream project—*star in a jar.*

Star in a jar

If nuclear fusion is the process that drives energy production in stars, it's quite natural to think about imitating such a process on Earth. For a long time, scientists have dreamed of the abundant clean energy that fusion could provide. There are many ongoing attempts in this regard, but the one that many scientists think could bring fusion energy to us is the International Thermonuclear Experimental Reactor (ITER). This is one of the most ambitious energy projects in the world

Fig. 8.5. *ITER tokomak. The world has invested in nuclear fusion as a future source of energy. Image courtesy of ITER.*

today—a carbon-free source of energy based on the same principle that powers the stars.

Located in southern France, ITER is a collaboration by 35 nations to build the world's largest tokamak (Fig. 8.5), a magnetic fusion device that has been designed to prove the feasibility of fusion on a large scale.

Earlier, we have seen how the mass-energy conversion can be calculated using the $E = mc^2$ equation. In stars, at the extremely high temperatures that exist there, hydrogen atoms fuse together to form the next higher element, helium. At these very high temperatures, hydrogen gas, in fact, becomes plasma—an ionized state of matter in which the negatively charged electrons are separated from the positively charged atomic nuclei (ions). During this fusion process, mass is converted into energy—once again, $E = mc^2$.

Due to the strong repulsive electrostatic force that exists between positively charged nuclei, the nuclear fusion is usually quite hard to begin. This force prevents the nuclei from getting close to each other, but at extremely high temperatures, they can overcome this force and the fusion process can be started. The temperatures and conditions present in stars are suitable for this to happen since the ions can move faster and come closer. The resulting fusion is accompanied by a tremendous amount of energy (Fig. 8.6). So, if we ever want to imitate this process on Earth, as we have been trying to do at facilities like ITER, we need to learn how to put a star in a jar.

The production of energy from mass is well understood. It changed our world when this idea was put into practice as a

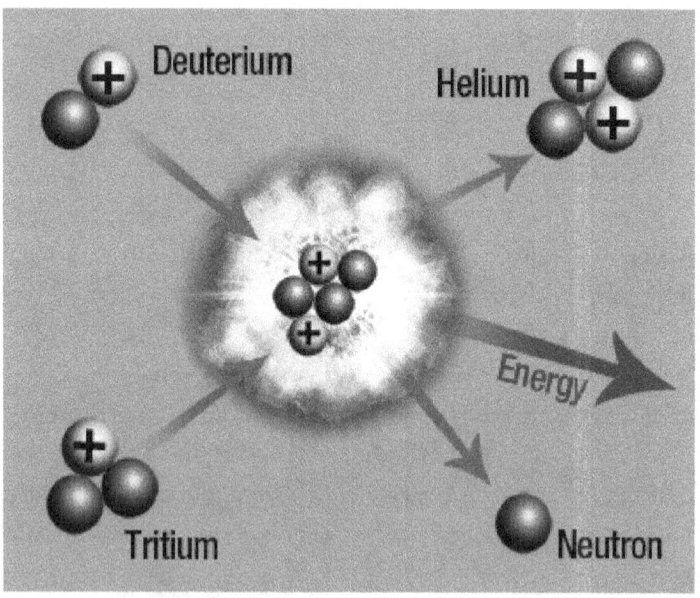

Fig. 8.6. *The fusion of deuterium and tritium, two isotopes of hydrogen. Image courtesy of Wikimedia Commons.*

weapon during the Second World War. Nuclear fission is the process in which a heavy element, like uranium or plutonium, undergoes fission to produce a huge amount of energy. So, if mass can become energy, a natural question that we might ask is: Can energy become mass?

Can energy turn into mass?

All the previous discussions we have had about the equation $E = mc^2$ have dealt with energy from mass whether it is fission, fusion, or the way particle accelerators operate. We can naturally think about the same equation in the reverse form and ask: Can energy turn into mass?

Quantum electrodynamics (QED) is a quantum field theory that deals with the electromagnetic field and its interaction with electrically charged particles. The ability to create matter from light is among the most striking predictions of quantum electrodynamics. In 1934, physicists John Wheeler and Gregory Breit predicted such a process and the simplest mechanism by which pure light could be potentially transformed into matter (Fig. 8.7).

In this process, the collision of two photons creates a positron-electron pair. (A positron is the antiparticle of an electron, essentially an electron with a positive charge.) The pair production was first observed in 1932, which led to two early Nobel Prizes in physics to Carl Anderson for the discovery of positrons (1936) and to Paul Dirac for the theory of antiparticles (1933).

Because the theory behind the Breit-Wheeler process was well understood, physicists were able to devise a practical method in which particles from energy could be created. However, in 2014, a breakthrough proposal was presented by

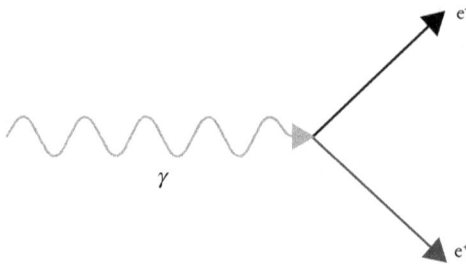

Fig. 8.7. *Creation of an electron and a positron by pair production. Image courtesy of Wikimedia Commons.*

physicists at Imperial College in London. They (Pike et al., 2014) offered a relatively simple way to physically demonstrate the Breit-Wheeler process. According to their proposal, an extremely powerful high-intensity laser beam would speed up electrons to just below the speed of light. These electrons would then be fired into a slab of gold to create a beam of photons a billion times more energetic than those of visible light. The next stage of the experiment involved a tiny gold can called a hohlraum (German for "empty room"). Scientists would fire a high-energy laser at the inner surface of this gold can to create a thermal radiation field. They would then direct the photon beam from the first stage of the experiment through the center of the can, causing the photons from the two sources to collide and form electrons and positrons.

Mass and energy were considered two separate things before Einstein's equation came into being. Einstein proved that neither mass nor energy was conserved separately but that they could be traded one for the other, and only the total "mass-energy" was conserved. It may seem to many that the idea of mass-energy, similar to the idea of wave-particle, is filled with paradoxes. However, according to physicist Richard Feynman, "paradox is only a conflict between reality and your feeling of what reality ought to be."

Fusion, fission, and pair production all make sense in light of the formula $E = mc^2$. The formula also tells us mass and energy are not different but are the same entity manifested in two different ways. Paradoxes may persist or vanish, but the mass-energy relation is forever.

Bibliography

Appell, D. (2013, October 15). The supercollider that never was. *Scientific American*. Retrieved from https://www.scientificamerican.com/article/the-supercollider-that-never-was/

Ashtekar, A., Berger, B., Isenberg, J. A., & MacCallum, M. A. (2015). *General relativity and gravitation: A centennial perspective.* Cambridge: Cambridge University Press.

Bernstein, J. (1991). *Einstein.* London: Fontana Press.

Bodanis, D. (2000). *E = mc²: A biography of the world's most famous equation.* New York: Walker & Co.

Born, M. (1962). *Einstein's Theory of Relativity.* New York: Dover.

Darrigol, O. (2000). *Electrodynamics from Ampère to Einstein.* Oxford: Oxford University Press.

Einstein, A. (1946). An elementary derivation of the equivalence of mass and energy. In Einstein, A. *Albert Einstein: Out of my later years* (pp. 116–119). New York: Bonanza, 1990.

Einstein, A. (1949). Autobiographical Notes. In Schilpp, P. A. (ed.), *Albert Einstein: Philosopher-Scientist.* The Library of Living Philosophers. La Salle, IL: Open Court.

Einstein, A. (1952). Relativity: The Special and the General Theory. New York: Three Rivers Press.

Einstein, A. (n.d.). Volume 2: The Swiss Years: Writings, 1900–1909 (English translation supplement) p. 172. Retrieved from https://einsteinpapers.press.princeton.edu/vol2-trans/186

Einstein, A., & Lawson, R. W. (1921). *Relativity: The special and general theory*. New York: Holt.

Galison, P. (2003). *Einstein's Clocks, Poincaré's Maps: Empires of Time*. New York: W. W. Norton.

Graneau, P., & Graneau, N. (2006). Albert Einstein – Inertia obscured by gravitation. In *In the grip of the distant universe* (pp. 166–188). doi:10.1142/9789812773807_0008

Graneau, P., & Graneau, N. (2006). *In the grip of the distant universe: The science of inertia*. Hackensack, NJ: World Scientific.

Lederman, L. M., & Teresi, D. (2006). *The God particle: If the universe is the answer, what is the question?* Mariner Books.

Lorentz, H. A. (1904). Electromagnetic phenomena in a system moving with any velocity less than that of light, *Proceedings of the Royal Netherlands Academy of Arts and Science, 6*, Amsterdam (reprinted, pp. 11–34).

Lorentz, H. A., et al., Perrett, W., & Jeffrey, G. B. (trans.). (1923 & 1952). *The Principle of Relativity*. London: Methuen & New York: Dover.

Mornas, L., & Alonso, J. D. (2006). *A Century of relativity physics: ERE 2005, XXVIII Spanish Relativity Meeting*, Oviedo, Asturias, Spain, 6–10 September 2005. Melville, NY: American Institute of Physics.

Pike O. J., et al. (2014). A photon–photon collider in a vacuum hohlraum. *Nature Photonics, 8*, (pp. 434–436). Retrieved from www.nature.com/articles/nphoton.2014.95.

Scientists discover how to turn light into matter after 80-year quest. (2014, May 18). Imperial College London. Retrieved from http://www.sciencedaily.com/releases/2014/05/140518164244.htm

Stephani, H., Kramer, D., MacCallum, M.A., Hoenselaers, C., & Herlt, E. (2009). *Exact solutions of Einstein's field equations*. Cambridge, UK: Cambridge University Press.

The Nobel Prize in Physics 1936. (n.d.). *Nobel Prize*. Retrieved from https://www.nobelprize.org/prizes/physics/1936/summary/

CHAPTER 9

The equation that counts the alien civilizations

Two possibilities exist: Either we are alone in the Universe or
we are not. Both are equally terrifying.
—*Arthur C. Clarke* quoted in *Visions: How Science Will
Revolutionize the Twenty-First Century (1999)*

Abstract

*Dr. Frank Drake presented an equation in 1961 that enables us to
estimate the number of technological civilizations that might exist in
our universe. Although it does not provide any particular solution, it
offers a way to identify factors that play a role in the development of
such civilizations. As we search for extraterrestrial intelligence, we cannot ignore the significance of this equation.*

About two decades ago, it was quite uncertain whether stars
other than our own Sun even hosted their own planets. However,

today, according to NASA, the latest count of confirmed exoplanets stands at around 4,000. A few of them are considered to be potential earths, although to reach such a conclusion we need more details about these planets. This count will definitely go up, and many researchers believe we will come in contact with intelligent life forms. With the advancement of technology, it is just a question of when.

Understandably, these discoveries will kindle questions about the Earth's place in the universe and will mark a major turning point in human history. While technological advancement is a key factor in the discovery of some form of life on another planet, the Drake equation set an agenda for the search for extraterrestrial intelligence. The formula, devised by astronomer Frank Drake in 1961, remains a fundamental equation as we search for extraterrestrial intelligence.

Frank Drake, a Harvard-trained radio astronomer, joined the National Radio Astronomy Observatory (NRAO), which was founded in 1956 and funded by the National Science Foundation (NSF). This observatory provides state-of-the-art radio telescope facilities for use by the international scientific community. At the NRAO facility in Green Bank, West Virginia, Drake set up radio telescopes in the Search for Extraterrestrial Intelligence (SETI), dubbed Project Ozma. He used the 85-foot Tatel telescope to observe the stars Tau Ceti and Epsilon Eridani for signs of civilization but with no apparent success.

However, in 1961, Drake came up with an equation (Fig. 9.1) and presented it at a meeting that was called to discuss the future course of SETI. Carl Sagan, who later gave SETI a well-known public face, attended the meeting along with Melvin Calvin who won the 1961 Nobel Prize in chemistry. The equation is the product of a number of parameters that give an estimated number of intelligent civilizations in our galaxy:

$$N = R_* * f_p * n_e * f_l * f_i * f_c * L$$

where

N = The number of civilizations in the Milky Way Galaxy whose electromagnetic emissions are detectable.

R_* = The rate of formation of stars suitable for the development of intelligent life.

f_p = The fraction of those stars with planetary systems.

n_e = The number of planets, per solar system, with an environment suitable for life.

f_l = The fraction of suitable planets on which life actually appears.

f_i = The fraction of life-bearing planets on which intelligent life emerges.

f_c = The fraction of civilizations that develop a technology that releases detectable signs of their existence into space.

L = The length of time such civilizations release detectable signals into space.

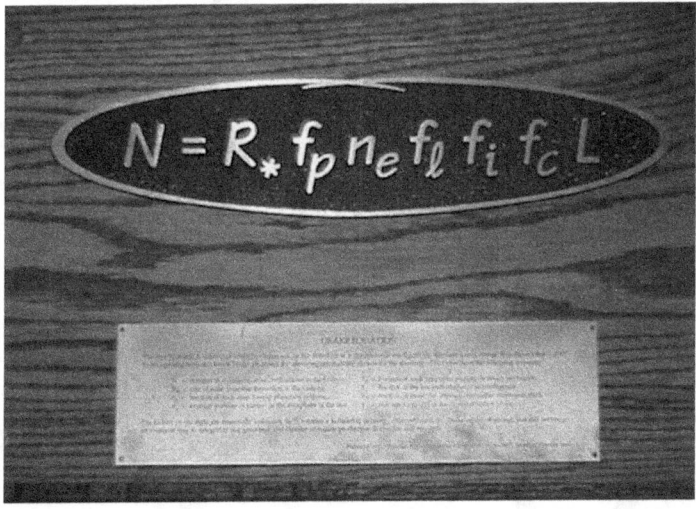

Fig. 9.1. *Drake equation plaque in the NRAO Green Bank Residence Hall Lounge. Note the smudges from so many people touching the plaque. Image courtesy of NRAO/NSF/AUI.*

So, what does this equation really mean? For now, there is no exact solution to this equation as several variables indicated can vary based on the data we receive. Look at the variables in detail in the following figure (Fig. 9.2). The difficulty in reaching a precise solution to this equation is compounded by the fact that most of the indicated variables are subjective, and their value can vary as we continue to explore the cosmos. For example, the estimation for fraction of stars with planets has definitely gone up as technology has enabled us to locate more planets outside our solar system in recent years.

At this point, we cannot simply put a number on other civilizations, but it is quite possible that we are not the only civilization in this universe. This argument is not based only on the vastness of our universe, which is true, but also on the number of different exoplanets we have come across in the last two decades. It is fair to assume the existence of extraterrestrial life as a possibility. Many SETI researchers believe that discovery is bound to happen even though we cannot put a timeframe on finding a civilization similar to an advanced one like ours. It may not happen any time soon. If there are a million technical civilizations in the Milky Way galaxy, the average separation between civilizations will be about 300 light-years.

Fig. 9.2. *The Drake equation representing each symbol of the equation. Image courtesy of Danielle Futselaar/SETI Institute.*

This explains the practical difficulty in making contact with one of them.

Nonetheless, this equation remains a powerful tool to guide us through the vast universe. We may never know the exact number of advanced civilizations that exist or that have existed in the past. They could have been wiped out completely for various reasons, including self-destruction. Yet, if we ever come across another civilization besides ours, that would mark a huge turning point in human history.

Impact of alien contact on human society

Contact with intelligent life elsewhere in the universe will shake up many of our existing notions as it will present not only scientific questions but also theological and philosophical conundrums. As you can imagine, many religions will find such an outcome deeply challenging. This is especially true for Christianity, which primarily focuses on humankind and teaches us that God created man in his own image and that all other animals and plants were created for mankind.

The idea of infinite space with the infinite glory of God originated with Nicholas of Cusa, a German philosopher who kept his infinite theology within the Catholic framework. But, as our explorations became more scientific, such thoughts gave way to practical astronomy. For example, in 2017, three scientists in the field of exoplanet science were named in *Time* magazine's annual list of the 100 most influential people.

The core question I would like to present here is: Does God's creation extend beyond a single planet? If so, would the inhabitants of those planets believe in the same Gods as humans do? How could the creator of the universe deny the inhabitants of those worlds a chance to redeem their sins? Does that mean that God incarnated as Jesus in those worlds contrary to Bible teachings that say that the redemption in Christ was a unique event meant for humans on Earth?

You could make a very strong case for institutional religions surviving the discovery of alien planets with intelligent inhabitants and the ensuing tussle with exotheology, a term that describes theological issues as related to extraterrestrial intelligence. These institutions have always shown an amazing ability to remain relevant. Whenever they encounter a new paradigm shift, they come up with interpretations from scriptures that justify their own existence. There is also, quite simply, something special about religion that resonates with humans on a fundamental level.

For traditional religions and religious institutions, the desire to expand their material wealth and power has often taken precedence over the spreading of theological doctrines. This has often led to a culture of exploitation of both people and the planet. This perhaps explain why the Copernican revolution or Darwinism didn't displace the religious order in a significant way in the past. The religious elite on Earth will find the courage and determination to pursue their goals in this material world even if they are convinced of the existence of multiple universes that operate under different laws of physics.

We know how our vision of the cosmos changed when we accepted the Copernican model that replaced the Ptolemaic model (the geocentric model that placed the Earth at the center of the universe). A major paradigm shift occurred when Copernicus established the heliocentric model with the Sun at the center of the solar system. The distinctiveness of our solar system

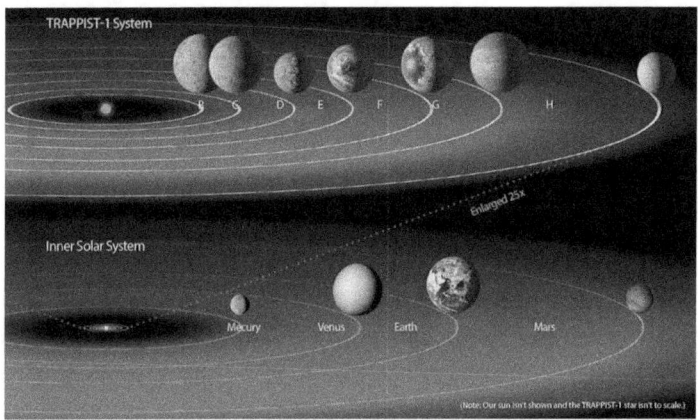

Fig. 9.3. *The TRAPPIST-1 system contains seven known Earth-sized planets. Three of them—TRAPPIST-1E, F and G—are located in the habitable zone of the star (shown in green in this artist's impression) where temperatures are just right for liquid water to exist on the surface. As a comparison to the TRAPPIST-1 system, the inner part of the solar system and its habitable zone is shown. Image courtesy of NASA/JPL-Caltech.*

as a unique system is vanishing quickly as astronomers identify similar systems outside our own. For example, TRAPPIST-1 is one of the exosolar systems (found outside our own solar system) which is about 39 light-years from Earth (Fig. 9.3). In 2017, astronomers identified four more planets in this system, making a total of seven planets orbiting the star.

In recent times, the Catholic Church has indicated that the discovery of intelligent life outside our system does not conflict with its faith. As the Church has pointed out, denying such existence would be like putting limits on the creative freedom of God. In 2008, the Vatican's chief astronomer in an interview titled, "The extraterrestrial is my brother," suggested extraterrestrial life (EL) could be part of the grand creation of God. The study, "Religious Understandings of Science (RUS)," by Rice

University in 2014, found that nearly 50% of evangelicals in the United States believe that science and religion can work together and support each other. More than ever before, there's room for debate about how religious communities should respond to scientific conclusions. When guidance can't be found in the Bible or other religious texts, it becomes a question of interpretation for individual believers. Buddhists might be the most comfortable with life on other planets, as their religious texts offer no limit to the size of the universe.

The triumph of these institutions is analogous to the audacity of organisms when facing challenges in nature. Religious institutions possess impressive survival skills, greater than individual human abilities. They also have their origins in personal and emotional needs and will continue to meet those needs.

Yet, this personalized religion became a complex organism with ambitions to control society and wealth. A recent Pew Research Center study found that the growth of major religious groups will outpace the rise in the religiously unaffiliated population, despite trends in the United States and other western countries where the proportion of religiously unaffiliated people is expected to grow.

So, how can we resolve the problem of many worlds and their many Gods? We can be certain that earthly religions will not accommodate the alien Gods. Perhaps we should turn to the astronomer Carl Sagan who wrote in *Cosmos*: "Meanwhile, elsewhere there are an infinite number of other universes each with its own God dreaming the cosmic dream... It is said that men may not be the dreams of the Gods but rather that the Gods are the dreams of men."

In the meantime, we should talk about the first-ever alien object that appeared in our solar system.

An alien messenger

On October 19, 2017, the first known interstellar object to visit our solar system was discovered by the University of Hawaii's Pan-STARRS1 telescope. This telescope is part of NASA's Near-Earth Object Observations (NEOO) Program, which finds and tracks asteroids and comets in Earth's neighborhood. The object was officially cataloged as 1I/2017 U1 ("1" for first and "I" for interstellar). It was named, "'Oumuamua," which is Hawaiian for "a messenger from afar arriving first." It was briefly classified as an asteroid, but later, when it was found to be accelerating, some thought it was a comet (Fig. 9.4). The truth is that this object didn't fit into the classic definition of asteroid or comet and, as such, confusion persisted about its identity. However, observations revealed no signs of cometary activity after 'Oumuamua went past the Sun at a blazing speed of 196,000 miles per hour (87.3 kilometers per second). When a comet moves along its path around the Sun, it will travel fastest when it is close to the Sun (perihelion) and will slow down when it is farthest from Sun (aphelion).

When more data came in, a more precise trajectory was determined, and it was found that this object wasn't slowing down as one would expect. This meant that some force was acting on it, not just the acceleration due to Sun's gravity, accelerating it very slightly. One thing everyone agreed on was the weirdness of this object as the shape of its orbit indicated it was not orbiting the Sun like a comet or asteroid. Furthermore, the brightness of this object changed by a factor of ten when it tumbled around within a period of eight hours as opposed to traditional asteroids and comets that change their brightness by a factor of three or less.

In 2018, in a paper titled "Could solar radiation pressure explain 'Oumuamua's peculiar acceleration?", Harvard University Professor and Chair of the Astronomy Department, Abraham Loeb, and Shmuel Bialy, a postdoctoral scholar at the Harvard Smithsonian Center for Astrophysics, presented an interesting scenario that 'Oumuamua may be a fully operational probe sent to our solar system by an alien civilization, perhaps intentionally. The authors of the paper put forward the possibility that 'Oumuamua might have been propelled by technology somewhat similar to a light sail or solar sail in which solar radiation propels the system. The paper did not make the definite

Fig. 9.4. *This artist's impression shows the first interstellar asteroid: 'Oumuamua. This unique object was discovered on 19 October 2017 by the Pan-STARRS1 telescope in Hawaii. Subsequent observations from ESO's Very Large Telescope in Chile and other observatories around the world show that it was traveling through space for millions of years before its chance encounter with our star system. 'Oumuamua seems to be a dark red, highly-elongated metallic or rocky object, about 400 meters long, and is unlike anything normally found in the solar system. Image courtesy of ESO/M. Kornmesser.*

claim that it is an alien craft but proposed such a possibility, and many discussions ensued in the scientific community.

At this point, we don't know for sure whether 'Oumuamua was propelled by gravity or alien technology, but it propels our curiosity. This is what Professor Loeb expressed in media interviews: that the possibility should be discussed. And 'Oumuamua was truly the very first alien object that came to our solar system since we began observing the asteroids and other objects that regularly pass close to the Earth's orbit. In his 2018 *Scientific American* article, titled "6 Strange Facts about the Interstellar Visitor 'Oumuamua," Professor Loeb observed that "in contemplating the possibility of an artificial origin, we should keep in mind what Sherlock Holmes said: When you have excluded the impossible, whatever remains, however improbable, must be the truth."

Professor Loeb's paper and subsequent opinions lay out a sincere scientific approach we should follow as we pursue space science with determination and a new focus. Our age-old quest for other worlds must take multiple pathways of search, and every possibility should be examined. We need to analyze every piece of information in light of the fact that NASA's Kepler mission revealed there are more planets than stars in our galaxy. The search for diverse worlds and objects warrants diverse approaches like solar sail technology. We may come across something entirely unlike what we have ever found in our solar system. In other words, the detection of radio waves alone should not be our sole approach in the search for extraterrestrial signatures.

While 'Oumuamua has been confirmed as an alien object, most of our thoughts about intelligent aliens are still on the fringes of speculative science. Additionally, even if the majority of alien civilizations turns out to be biological, it may be that the most intelligent alien civilizations will be ones in which the inhabitants are super intelligent and have no interest in the resources from a tiny planet like ours. They would probably harness their resources from beyond a planet or star. Even for technologically advanced civilizations, interstellar voyages would probably be justified only for major purposes, and plundering the Earth for its resources would be neither practical nor desirable. But we will continue to search for them because, as Carl Sagan remarked, "In the deepest sense, the search for extraterrestrial intelligence is a search for ourselves." Because the question, "What is our place in the universe?", might just be genetically encoded in our species.

In the age of Messaging Extraterrestrial Intelligence (METI), who should speak for Earth?

For decades, astronomers have been searching for a message from other civilizations in the universe. In the forefront of such efforts are the organizations like SETI that analyze the radio signals coming from space in order to locate possible intelligent alien civilizations. Now, a group of scientists are trying something different, a proactive form of search—Messaging Extraterrestrial Intelligence (METI).

In 2017, a group of scientists, associated with METI, and artists from Sónar, Barcelona's music, creativity and technology festival, joined together to beam a message towards a red dwarf star (GJ 273) that harbors a planet in its habitable zone about 12.4 light-years away from Earth. If any aliens happened to receive this message and decided to respond the same way using signals, we would receive their response in about twenty-five years. (The time signals traveling at the speed of light will need to cover the back-and-forth distance from Earth.)

This was not only a celebration of an artistic and scientific experiment but also a collective rumination on what it means to be human and alien. In an effort to make the first human contact with extraterrestrial intelligence, participants chose GJ 273 as the target star, also known as Luyten's star after Dutch-American astronomer Willem Jacob Luyten. In March 2017, it was discovered that GJ 273 had two planets. One of them, known as GJ 273b, located in the habitable zone, could potentially harbor liquid water, and perhaps life.

The METI alien hunters used the EISCAT 930 MHz transmitter located in Ramfjordmoen, near Tromsø, Norway. They encoded the music-based message that involved mathematics in a binary system and beamed it in the direction of GJ 273b.

Even this type of intentional message sent to the planet GJ 273b is, to an extent, a symbolic measure as there is a huge uncertainty about the possibility of any civilizations receiving or even responding to these messages. There is no convincing evidence that this planet hosts intelligent life though it's located within the habitable zone of the red dwarf star.

At a time when many scientists are confident that the detection of extraterrestrial life is more a question of when than if, this act enraged some astronomers as they pointed out that inviting aliens would pose an existential risk for humanity. Setting aside the possible dangerous outcomes stemming from contacting aliens, an important question remains for all of us to answer: *Who should speak for Earth?*

Humans have been transmitting signals into space for decades. The high-frequency radio signals we rely on for radio and television may already have reached thousands of nearby stars. Any technologically advanced civilization with a sensitive antenna can listen to those signals if it exists within the range of those transmissions. In 2008, NASA beamed a song—the Beatles' "Across the Universe"—directly into deep space to commemorate the 40th anniversary of the day the Beatles recorded the song, as well as the 50th anniversary of NASA's founding.

Currently, there are no laws or regulations in regards to sending messages across the universe. Anyone with a powerful transmitter can send messages to alien worlds. Considerable thought should be given to having a plan in place for humans to communicate with aliens. In the absence of such a plan, individual organizations or individuals themselves will fill the void and speak for Earth—and that could be disastrous.

We have seen how we deal with the problems we create for ourselves, such as climate change, that could significantly disrupt future life on this planet. Political leaders, individual nations and corporations have stood for their narrow objectives and have collectively forgotten to stand for this planet. It should not happen again. The voices of the citizens of this planet should be heard before formalizing a plan to communicate with aliens. Individual organizations or countries alone should not be allowed to dictate the terms of communication with aliens, whether that happens in the future or never. We are children of the Earth and sky, and we should have a voice in the affairs of both.

While our search for intelligent alien life is a long shot, recent confirmation about the existence of organic molecules on Mars shows the possibility of widespread life in the universe. In 2018, it was reported that NASA's Curiosity rover (Fig. 9.5) discovered ancient organic molecules on Mars, embedded within sedimentary rocks that are billions of years old. This finding also suggests the planet could have supported ancient life. The Curiosity rover

Fig. 9.5. *NASA's Curiosity rover has discovered ancient organic molecules on Mars, embedded within sedimentary rocks that are billions of years old. Image courtesy of NASA/GSFC.*

has been taking samples on Mars since its arrival in 2012. It has also recorded atmospheric measurements of methane on Mars with variation levels linked to seasons. These latest findings once again open up the question: Is life commonplace throughout the universe?

In a galaxy that is 100 light-years across and filled with billions of stars and planets, it is only a question of time before we come across intelligent life, unless our planet is so unique that such life exists only on Earth. For the time being, we must be content with what science fiction writer Arthur C. Clarke once remarked, "I'm sure the universe is full of intelligent life. It's just been too intelligent to come here."

As technology advances, the search for extraterrestrial intelligence will accelerate and may yield results that we can't even imagine now. The Drake equation may need to be refined, but its significance will not fade even after we find intelligent aliens. After all, this equation opened the door to the discovery of other intelligent beings and reflects our desire to be part of a world without borders.

Bibliography

Anatomy of an exosolar system. (n.d.). National Aeronautics and Space Administration. Retrieved from https://asd.gsfc.nasa.gov/blueshift/index.php/2012/08/03/jillians-blog-anatomy-of-an-exosolar-system/

Believing in aliens not opposed to Christianity, Vatican's top astronomer says. (2008, May 13). *Catholic News Agency*. Retrieved from https://www.catholicnewsagency.com/news/believing_in_aliens_not_opposed_to_christianity_vaticans_top_astronomer_says

Crowe, M. J. (1986). *The extraterrestrial life debate, 1750–1900*. Cambridge University Press.

Davies, P. (2013, November 18). Are We Alone in the Universe? *The New York Times*. Retrieved from https://www.nytimes.com/2013/11/19/opinion/are-we-alone-in-the-universe.html

Dick, S. J. (1996). *The biological universe: The twentieth century extraterrestrial life debate and the limits of science*. New York: Cambridge University Press.

Dick, S. J. (2015). Introduction. In Vakoch, D. A., & Dowd, M. F. (eds.), *The Drake Equation: Estimating the prevalence of extraterrestrial life through the ages* (Cambridge Astrobiology, pp. 1–20). Cambridge: Cambridge University Press. doi:10.1017/CBO9781139683593.003

Drake, F. (1959). How can we detect radio transmissions from distant planetary systems? *Sky and Telescope, 19*, 140. Repr. in Cameron 1963a, 165–175.

Drake, F. (2013, June 20). Reflections on the Equation. Special Issue on the Drake Equation. *International Journal of Astrobiology, 12*(3), 173–176.

Ecklund, E. H. (2014). Religious communities, science, scientists, and perceptions: A comprehensive survey. Paper presented at the

Annual Meetings of the American Association for the Advancement of Science. Retrieved from https://www.aaas.org/sites/default/files/content_files/RU_AAASPresentationNotes_2014_0219%20(1).pdf

Eigenbrode, L. J., et al. (2018, June 8). Organic matter preserved in 3-billion-year-old mudstones at Gale crater, Mars. *Science, 360*(6393), 1096–1101. doi: 10.1126/science.aas9185. Retrieved from http://science.sciencemag.org/content/360/6393/1096

Kaku, M. (1999). *Visions: How science will revolutionize the twenty-first century*. Oxford: Oxford University Press.

Kepler. (n.d.). Jet Propulsion Laboratory, California Institute of Technology. Retrieved from https://www.jpl.nasa.gov/missions/kepler/

Loeb, A. (2018, September 27). How to search for dead cosmic civilizations. *Scientific American*. Retrieved from https://blogs.scientificamerican.com/observations/how-to-search-for-dead-cosmic-civilizations/

Loeb, A. (2018, November 20). 6 Strange facts about the interstellar visitor 'Oumuamua. *Scientific American*. Retrieved from https://blogs.scientificamerican.com/observations/6-strange-facts-about-the-interstellar-visitor-oumuamua/

Loeb, A, & Bialy, S. (2018, November 12). Could solar radiation pressure explain 'Oumuamua's peculiar acceleration? *The Astrophysical Journal Letters, 868*(1). Retrieved from http://iopscience.iop.org/article/10.3847/2041-8213/aaeda8/pdf

Mathew, S. (2017, April 26). The discovery of alien life may be close. How will religion survive it? *The Guardian*. Retrieved from https://www.theguardian.com/commentisfree/2017/apr/26/discovery-of-alien-life-religion-will-survive

METI International. (n.d.). Retrieved from http://meti.org/

NASA finds ancient organic material, mysterious methane on Mars. (2018, June 7). Jet Propulsion Laboratory, California Institute of Technology. Retrieved from https://www.jpl.nasa.gov/news/news.php?feature=7154

National Radio Astronomy Observatory. (n.d.). Retrieved from https://public.nrao.edu/

Pew Research Center. (2017, April 5). The changing global religious landscape. Retrieved from http://www.pewforum.org/2017/04/05/the-changing-global-religious-landscape/

Scoles, S. & Heatherly, A. (2011). The Drake equation: 50 Years of giving direction to the scientific search for life beyond Earth. *The Universe in the Classroom*. Retrieved from https://astrosociety.org/edu/publications/tnl/77/77.html

Seth, S. (January, 2019). *The Drake Equation: Could It Be Wrong?* Retrieved from https://www.seti.org/drake-equation-could-it-be-wrong

The 100 Most Influential People. (2017). *Time*. Retrieved from http://time.com/collection/2017-time-100/

Wall, M. (2017, November 16). Interstellar message beamed to nearby exoplanet. *Scientific American*. Retrieved from https://www.scientificamerican.com/article/interstellar-message-beamed-to-nearby-exoplanet/

CHAPTER 10

The equation that tells there is order in disorder

If the flap of a butterfly's wings can be instrumental in
generating a tornado, it can equally well be instrumental in
preventing a tornado.
—*Edward Lorenz* presented in American Association for the
Advancement of Science (1972)

Abstract

We know gravity keeps planets in orbit and stars in galaxies and galaxies in the universe. Yet, how do all these keep going like clockwork? Did the universe originate from chaos? If so, how do we see order in disorder? The chaos theory suggests that time spent trying to predict the future of complex, nonlinear dynamic systems may be better spent elsewhere. Can we predict the future of our universe or, rather, should we look somewhere else? The paths of many asteroids and their orbits are chaotic. Can we use chaos theory to improve predictions in a complex system such as our solar system or our universe?

The term "butterfly effect" is associated with a situation where a very small change can produce a large effect over time. This term is attributed to Edward Lorenz, the MIT mathematician and meteorologist. He was also a pioneer of chaos theory which helped scientists describe a series of phenomena from the field of dynamics—the field of physics concerning the effect of forces on the motion of objects.

In 1972, Edward Lorenz delivered a speech entitled "Predictability: Does the Flap of a Butterfly's Wings in Brazil set off a Tornado in Texas?" Lorenz used this now-famous phrase to drive his ideas about chaotic systems, specifically the dynamics of atmosphere, which are extremely sensitive to initial conditions and initial parameters and may result in a completely different behavior of a complex system. The term "butterfly effect" stuck, and it has become a phrase easily embraced by popular culture. Several systems, from solar system to stock market, are potentially chaotic as huge changes in outcome may result from minor changes in initial conditions.

In linearity, the sum of the inputs will result in an output and is fairly simple mathematically. In other words, the whole is the sum of its parts. Nonlinear systems, in which the output of the system is not directly proportional to the input, are widespread in nature. Chaotic systems are nonlinear, dynamic, and deterministic in a purely mathematical sense. Thus, it may seem very simple to deal with them, yet they produce completely random behavior. The deterministic nature of these systems does not make them predictable. These systems are deterministic in the sense that the same input will produce the same output, but that does not mean they are predictable because the behavior of the system changes in a haphazard fashion. Furthermore, nonlinear systems are dynamic as they change over time depending on the current state. Linearity is a reductionist's dream, and nonlinearity can sometimes be a reductionist's nightmare (Mitchell, 2011).

The study of chaotic systems flourished with the advent of new and powerful computers. The mathematical representation of a dynamic system in a computer shows a figure, called an attractor, that reflects the movement of an object. A simple attractor can be a circle or an ellipse, but even a slight change in the system can cause it to become more unpredictable, as represented in Figure 10.1 below.

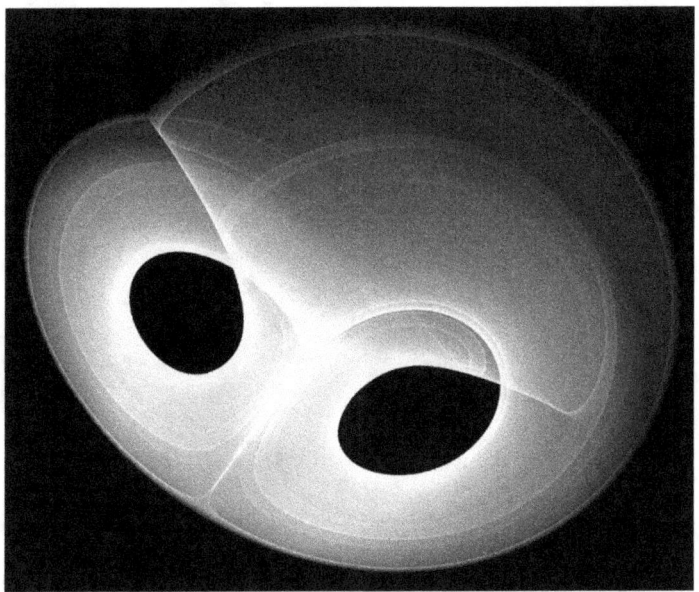

Fig. 10.1. *Lorenz attractor showing multiple pathways for a chaotic system. This indicates that a particle's speed and location are completely indeterminable even if we know the initial conditions. Image in the public domain via Wikimedia Commons.*

As a mathematician and meteorologist, Lorenz studied weather systems, which are classical examples of chaotic systems, at a time when forecasting was not even close to what we see today. He changed the initial conditions of a computer-generated weather system to assess the outcome and realized that huge uncertainty creeps in as the system evolves. This behavior and unpredictability of the system due to its sensitivity to initial conditions is what Lorenz expressed metaphorically as the butterfly effect—the simple flapping of a butterfly's wings can trigger devastating hurricanes thousands of miles away.

In order to study chaotic systems, scientists depend on something called the logistic map (also known as the logistic difference equation) which is essentially an equation that connects the future value with the initial value of a dynamic system. In fact, this equation was popularized by mathematical biologist Robert May (1976) when he modeled the behavior of different generations of a species. He noticed that a seemingly simple mathematical model could give way to complex dynamics. The basic form of the map is:

$$x_{t+1} = k\, x_t\, (1 - x_t)$$

where

x_t – initial value
k – constant that represents growth rate
x_{t+1} – future value

In his 1976 paper, May reported that such equations, even though simple and deterministic, could exhibit a surprising array of dynamic behavior, from stable points to a bifurcating hierarchy of stable cycles to apparently random fluctuations. There were, consequently, many fascinating problems, some concerned with the delicate mathematical aspects of the fine structure of the trajectories and some concerned with the practical implications and applications. The logistic map could be interpreted as a simple mathematical expression that considers the effects of feedback on future values of a system.

It should be mentioned here that Newton and many of his contemporaries envisioned a deterministic and predictable universe where events within that universe were also supposed to follow the laws of nature. However, the minor change in the initial condition of the simulation and the surprising outcome that Lorenz toyed with had a shocking impact for many in the scientific world. Now, it looks like our ability to predict future events is flawed, whether in fluid mechanics or weather forecasting. So, in the real world (most real-world systems are nonlinear dynamic), unpredictability is inherent in systems and events. We can, at best, make only a probabilistic conclusion.

The connection between chaos theory and various disciplines is obvious in the fields of population biology, epidemiology, genetics, social sciences, and economics. In addition, our desire to predict the weather more accurately makes it a perfect candidate to be considered under chaos theory. But, more importantly, the planetary motion and the origin of the universe itself could come under the purview of chaos theory.

Chaos and universe

Since ancient times, the motion of the planets around the Sun was observed and studied by humans with eagerness. The highly predictable orbits of the planets proposed by Johannes Kepler, and later reinforced by Newtonian mechanics, represent the prominent view of our cosmos in modern times—even today.

However, a radical shift in this direction arose with the idea of a chaotic solar system that was proposed by astronomer Jacques Laskar in a paper titled, "Large scale chaos and marginal stability in the solar system" (1996). His work provided evidence that large-scale chaos is present everywhere in the solar system and that it plays a major role in the sculpting of the asteroid belt and in the diffusion of comets from the outer region of the solar system. Moreover, all the inner planets probably experienced large-scale chaotic behavior for their obliquities during their history. Laskar's work also predicted that, due to the chaotic behavior of the orbits of the planets, Mercury could escape or collide with Venus in less than 3.5 billion years. The stability we witness now in the solar system thus seems to be of a marginal nature.

Another work that pointed to the chaos in the solar system came from researchers at MIT (Sussman & Wisdom, 1988) who used a program called Digital Orrery to perform an integration of the motion of the outer planets in our solar system over 845 million years. This integration indicated that the long-term motion of the planet Pluto was chaotic; its long-term shape depended very sensitively on the exact input parameters used to start the calculations. The imprecise location of Pluto, even around one kilometer today, could result in a range of possible orbits in the future, which is contrary to the deterministic view of Newtonian physics.

Furthermore, in recent years, astronomers, using a similar computer program, found that the orbits of the inner planets (Mercury, Venus, Earth, and Mars) may not be stable over the time scale of a few billion years. These results are not universally accepted, but they acknowledge the fact that the solar system, long considered to run like clockwork, is being questioned now by a few researchers.

It is generally believed that the Newtonian approach and Newton's laws still hold for our solar system, but there is growing recognition that a chaotic dynamic is ubiquitous in the solar system. It may sound shocking to entertain the notion that our solar system is chaotic in nature. However, the orbits of the planets could become chaotic, and the long-term impact of minor variations that we currently ignore could be a huge problem in the far future. The chaos theory teaches us that instability and disorder could arise from the systems we perceive as orderly and predictable now.

The chaotic view of the solar system goes against the traditional view whereby knowing the initial position of objects allows us to predict where they will be in the future. This idea was highlighted by French mathematician Pierre Laplace (1902) who wrote, in *A philosophical essay on probabilities*:

> We may regard the present state of the universe as the effect of its past and the cause of its future. An intellect which at a certain moment would know all forces that set nature in motion, and all positions of all items of which nature is composed, if this intellect were also vast enough to submit these data to analysis, it would embrace in a single formula the movements of the greatest bodies of the universe and those of the tiniest atom; for such an intellect nothing would be uncertain and the future just like the past would be present before its eyes.

Such notions could change in light of what we have learned from chaos theory. However, the time scale on which the solar system operates makes such a behavioral change unlikely in the near future.

Fig. 10.2. *Is our solar system a chaotic system? Some recent studies on geological features on Earth reject the classical clockwork mechanism of the solar system and point to a more chaotic system that will have huge impact on climate science. Image courtesy of Pixabay.*

It is surprising to note that other evidence for the chaotic nature of our solar system (Fig. 10.2) came from geological studies. Using evidence from alternating layers of limestone and shale laid down over millions of years in a shallow North American seaway, a team of researchers (Ma et al. 2017) discovered the 87-million-year-old signature of a "resonance transition" between Mars and Earth. During orbital motion, planets can come close to each other and thus exert a stronger gravitational force on each other. There can be a disruption in the resonance between Mars and Earth if the solar system is chaotic, and this can affect a planet's location and tilt, which in turn affects the amount of sunlight the planet receives. Since the amount of sunlight the Earth receives is linked to its climate, ultimately the disruption in the resonance can leave its signature on Earth. The researchers identified such effects on rocks they analyzed as part of this study.

This resonance disruption is the consequence of the "butterfly effect" in chaos theory that states small changes in the initial conditions of a nonlinear system can have large effects over time. These researchers presented geological evidence for a chaotic resonance transition associated with interactions between the orbits of Mars and Earth. Their analysis confirmed the predicted chaotic dynamic behavior of the solar system. The finding is significant as it provides the first hard proof for the "chaotic solar system" proposed earlier. In 2015, the Hubble telescope provided data that indicates Pluto's moons are "tumbling in absolute chaos"—one more piece of evidence to add to the ever-increasing notion that chaos is pervasive.

Fig. 10.3. *NASA's Spitzer and Hubble Space Telescopes have teamed up to expose the chaos that baby stars are creating 1,500 light-years away in a cosmic cloud called the Orion nebula. Image courtesy of NASA/ JPL-Caltech/STScI.*

If our solar system is chaotic, what about the universe itself? The order we see today has emerged from the chaos that existed in the past. So, what is fundamental order or disorder? To answer this question, we should peer into the heart of a nebula, which is an interstellar cloud of gas and dust thrown out by the explosion of a dying star such as a supernova. Nebulae also function as stellar nurseries where new stars begin to form. The Orion nebula, which is located 1,500 light-years away from Earth, is home to about 1,000 young stars (Fig. 10.3).

Perhaps it is time to rethink our perception of chaos. Our naïve view of the universe is based on cause and effect, and so are many of the physical theories that we rely on. So, it is somewhat uncomfortable to think that chaos, which is closely linked to disorder, is the root cause of this universe. Order and disorder are creations of our human mental faculty. We attempt to embrace beautiful and elegant solutions, but the reality may be different than what we imagine. Our universe may be engaged in a chaotic rhythm rather than a choreographed dance.

Chaos theory in modeling brain and climate change

In the twenty-first century, two of the most sought-after scientific models are brain mapping initiative and global climate change predictions. Although great progress has been achieved in both areas, they are still considered to be in their infancy. Given the inherent complex nature of these systems, researchers trying to get better models have turned to chaos theory.

Can we model our brain using our own brain? This seems like circular reasoning. Taming Artificial Neural Networks

(ANN) to mimic our own nervous system has received much attention these days. The application of chaos theory to neurobiology is a fairly new area of study, and some researchers believe that this approach will be helpful in better understanding the brain processes (Chay, 1996 & Rabinovich, 1998). Dynamic models of the brain developed by research reveal that the functional assembly of a brain is more than the sum of its parts. One way to find answers to questions about neural chaos is to build models that produce similar emergent chaotic behavior in artificial neural networks (Andras, 2003).

Assuming that the brain processes exhibit chaotic behavior, the proponents of this approach are trying to more accurately simulate brain activity both on the microscopic (neuron) level and the macroscopic (overall brain activity) level. The chaos "may be the chief property that makes the brain different from an artificial-intelligence machine" (Freeman, 1991). The functioning organs are generally considered to be linear and deterministic, but as dynamic systems, chaos theory could be a better fit to explain these functions. For example, the beats of a heart in a periodic fashion show the dynamic behavior of the system, in this case, the heart. Similarly, the brain of a living creature with all the actions going on is another example.

In chapter 6, we discussed how living organisms are capable of increasing their complexity without adding entropy as they organize orderly functions from disorder and exhibit intricate dynamic behavior. Now chaos theory, as a new approach, has shed light on a few aspects of organ functioning.

Climate change is one area where humankind is going to invest time and energy in the coming years. So, let's look at how chaos theory can play a role in that too. The key driver of climate change is the amount of solar radiation the Earth receives.

The greenhouse gases in the atmosphere create an imbalance between the solar radiation received and emitted, eventually warming the planet.

It has been predicted that climate change will generate stronger hurricanes as ocean temperatures rise. Does chaos theory offer hope to modify these hurricanes or, despite all our advances, will these giant systems remain out of our control? Hurricanes occur worldwide and are also known by other names, such as typhoons or tropical cyclones, in different parts of the world (Fig. 10.4). These large rotating systems of oceanic tropical origin maintain surface winds of at least 74 mph in a storm. Will these fearful forces of nature be forever beyond our control? Currently, we don't have many choices. It's fight or flight when we deal with these monsters, yet we keep trying to control them.

The most ambitious experiment to modify hurricanes was conducted between 1962 and 1983 in the United States. This project, named Stormfury, provided ambiguous results as scientists were unable to distinguish between the expected results of human intervention and the natural behavior of hurricanes. Lately, various ideas have been floated to modify hurricanes, including the deployment of an array of Earth-orbiting solar-power stations that could provide heat energy to change the temperature in and around a hurricane, which would cause it to shift its path in a predictable direction or slow its winds. Another prospective method discussed to modify these storms is to limit the availability of energy by coating the ocean surface with a thin film of a biodegradable oil that slows evaporation.

Scientists and engineers debate various techniques, from cloud seeding to cooling the oceans, to control the impact of climate change. Yet, regardless of method, we depend on computer models to analyze this pressing problem. When input

Fig. 10.4. *This GOES-13 Satellite imagery from June 25, 2010 shows the powerful Hurricane Celia (left), with the larger eye, and behind it is Hurricane Darby (right) with a much smaller eye. Image courtesy of NASA Goddard Space Flight Center, licensed under CC BY 2.0.*

parameters are tweaked, as Edward Lorenz demonstrated, the computer models generate highly unpredictable and random behavior. Lorenz, the father of chaos theory, described chaos as "when the present determines the future, but the approximate present does not approximately determine the future."

So, one could ask: How can climate be predictable if weather is chaotic? It is true that these models demonstrate the extreme sensitivity to initial conditions equivalent to the "butterfly" pattern seen in Figure 10.1. However, the existence of climate change in these models is visible as a common signature. In fact, Lorenz never claimed that chaotic behavior in a system made every prediction about that system impossible, as some have claimed.

What makes chaos theory a fundamental principle while we deal with the unknown? The atmospheric features and weather systems of the Earth are quite important to us, but such systems exist on other planets too. The reach of chaos theory is not limited to this planet or its climate, as you can see in the atmospheric features of Jupiter in Figure 10.5.

It is quite fair to assume that, in coming years, chaos theory will find its way to many of the untamed systems that exist, from social science to cosmology. Concealed in this idea is the ultimate liberator of human fear and a solution to the cry and despair that we hear around the world. For now, while these systems remain highly unpredictable, many researchers are optimistic about the way we can handle these systems. Recently, scientists have used machine learning—the same computational technique behind recent successes in artificial intelligence—to predict the future

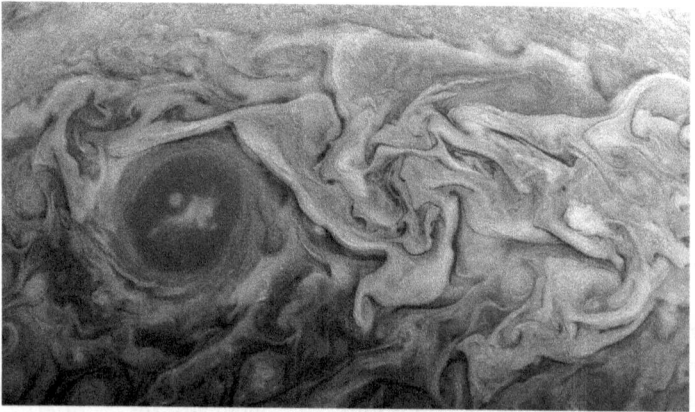

Fig. 10.5. *Dramatic atmospheric features in Jupiter's northern hemisphere are captured in this view from NASA's Juno spacecraft. The new perspective shows swirling clouds that surround a circular feature within a jet stream region called "Jet N6." Image courtesy of NASA/JPL-Caltech/SwRI/MSSS/ Kevin M. Gill.*

evolution of chaotic systems out to stunningly distant horizons (Pathak, J. et al., 2018). More progress will emerge in coming years, but chaos theory also teaches us there are, perhaps, limits to predictions.

So, the most famous equation in the science of dynamic systems and chaos tells us that order comes from disorder, or it can be seen as disorder emerging from order. Until we see the bottom of chaos theory, unpredictability will rule the Earth and the heavens, and it does not care about the wishes of our transient species.

Bibliography

Andras, P., & Lycett, S. (2007). An advantage of chaotic neural dynamics. IEEE International Conference on Neural Networks. Conference Proceedings, 1417–1422. doi: 10.1109/IJCNN.2007.4371166.

Bereiter, B., et al. (2018, January 4). Mean global ocean temperatures during the last glacial transition. *Nature, 553*, 39–44. doi:10.1038/nature25152.

Boeing, G. (2016). Visual analysis of nonlinear dynamical systems: Chaos, fractals, self-similarity and the limits of prediction. *Systems, 4*(4), 37–54.doi: 10.3390/systems4040037.

Chaos in the atmosphere. (2018 February). *The Discovery of Global Warming*. Retrieved from https://history.aip.org/climate/chaos.htm

Chay, T. R. (1996). Modeling slowly bursting neurons via calcium store and voltage-independent calcium current. *Neural Computation, 8*(5), 951–978.

Dizikes, P. (2011). When the Butterfly Effect Took Flight. *MIT Technology Review*. Retrieved from https://www.technologyreview.com/s/422809/when-the-butterfly-effect-took-flight/

Dvorak, R. (2006). The role of resonances in planetary systems. *International Journal of Bifurcation and Chaos, 16*(6), 1633–1644. doi: 10.1142/s021812740601557x.

Freeman, W. J. (1991). The physiology of perception. *Scientific American, 264*(2), 78–85.

Freeman, W. J., et al. (2001). Biocomplexity: adaptive behavior in complex stochastic dynamical systems. *Biosystems, 59*(2), 109–123.

Gross, D. (2008). The importance of chaos theory in the development of artificial neural systems. Retrieved from https://pdfs.semanticscholar.org/5956/f70e72a8056402e6d2a6d92fd37def7f5c8d.pdf

Kafka, F., & Brod, M. (1991). *The blue octavo notebooks*. Cambridge, MA: Exact Change.

Laplace, P. S. (1902). *A philosophical essay on probabilities*. New York: J. Wiley & Sons.

Laskar, J. (1988). Secular evolution of the solar system over 10 million years. *Astronomy and Astrophysics, 198*(1–2), 341–362.

Laskar, J. (1996). Large scale chaos and marginal stability in the solar system. *Chaos in Gravitational N-Body Systems*, 115–162. doi: 10.1007/978-94-009-0307-4_10.

Lorenz, E. N. (1972). Predictability: Does the flap of a butterfly's wings in Brazil set off a tornado in Texas. American Association for the Advancement of Science.

Lorenz, E. N. (1993). *The Essence of Chaos*. University of Washington Press.

Ma, C., et al. (2017). Theory of chaotic orbital variations confirmed by Cretaceous geological evidence. *Nature, 542*, 468–470.

Mandel, D. R. (1995). Chaos theory, sensitive dependence, and the logistic equation. *American Psychologist, 50*(2), 106–107.

May, R. (1976). Simple mathematical models with very complicated dynamics. *Nature, 261*, 459–467.

Mitchell, M. (2011). *Complexity: A guided tour*. Oxford: Oxford University Press.

Murray, N., et al. (1998). On the origin of chaos in the solar system. *Astron J, 116*, 2583–2589.

Pathak, J., et al. (2018). Model-free prediction of large spatiotemporally chaotic systems from data: A reservoir computing approach. *Physical Review Letters, 120*(2). doi: 10.1103/PhysRevLett. 120.024102.

Rabinovich, M. I., & Abarbanel, H. D. I. (1998). The role of chaos in neural systems. *Neuroscience, 87*(1), 5–14.

Sussman, G., & Wisdom, J. (1988). Numerical evidence that the motion of Pluto is chaotic. *Science, 241*(4864), 433–437. Retrieved from http://www.jstor.org/stable/1701411

Conclusion:
the death of equations

In this age of big data, many people think that it is time we abandoned traditional equations altogether in favor of data and algorithms. It's true that this approach might better serve us in short-term decision-making, but our greatest insights into nature's deep laws will continue to take the form of equations, and data and algorithm can't replace them. The equations have changed the world, and they will change it again for the generations to come. The equations may evolve and take new forms as we continue to search for new vistas, but the desire to connect various things and to produce a systematic result will never vanish.

Today, the many diverse areas of science and engineering struggle not due to a lack of data but due to the challenge of extracting the equation from the data. The models we generate from data often remain inconclusive. In the absence of a straightforward equation (as in climate science, neuroscience, epidemiology, etc.) to predict the future outcome, researchers rely on data, which often accumulates to a huge amount. Unfortunately, sometimes this data comes with inherent noise that destroys the true outcome.

To give just a hint of how equations and data are intertwined, let us look at the data-driven models of climate science. It should be emphasized here that, while ignoring a small amount of uncertainty, scientists agree that these models are accurate in predicting climate change. Yet, it is imperative that we collect more data if we want to improve the accuracy of these models in the future. We know that modeling the Earth's climate is one of our most daunting tasks. In order to generate a comprehensive model, we need to account for all interacting components, such as atmosphere, clouds, moisture, and oceans, to list a few. The models we see today are a simulation of all the known factors that can affect Earth's climate. While some of these factors do not change (elevation and latitude), others vary constantly (seasons, volcanic eruptions, air pollution, etc.). In the real world, all of these things are connected, and researchers try to link these parameters using algorithmic simulations to create successful models.

Now, imagine that we have an equation that can predict climate change with all the required accuracy. That equation would be very precise, and we should be able to come up with clear-cut values for climate change. But we do not have a single equation to do that job—at least for now. This in no way diminishes the conclusions of scientists who agree on climate change, and we cannot wait for an equation to mitigate or adapt climate change. This is an example of a case of where the absence of a working equation makes the field so complicated; we have a huge amount of data accumulation, yet the outcome is not precise. We have seen in chapter 10 how even a deterministic equation can yield unpredictable results, and that explains the practical difficulty in dealing with systems like climate.

In the classical physics era, the discovery of equations was propelled by curious minds who carefully observed the behavior and expression of nature around them. Today, the extraction of a governing equation from data is a more serious challenge since the method of least squares or regression analysis does not always result in equations. This is partly because of the fact that there is no guarantee that all data comes with rich information; it could be filled with noise. To get a sense of the amount of data we generate now and in the past, see what Google's Eric Schmidt said in 2010: "There were five exabytes of information created between the dawn of civilization through 2003, but that much information is now created every two days." Remember 1 exabyte = 10^{18} bytes, and this data generation rate is increasing exponentially.

With this kind of abundant data, we might hope that we could tame the dynamic systems. Unfortunately, the laws that govern some of these systems remain elusive, and data-driven discovery of dynamics will continue to play an important role in these efforts. The fields that heavily depend upon a data-driven approach, such as brain mapping or inferring climate change, will continue to suffer from a common infirmity—the lack of an equation.

The absence of a credible equation to control a system in many fields is creating more data and debate. In fact, as a society, we are enslaved by the culture of data. In my view, equations can function as liberators from this data enslavement. The need for a colossal data center and the supreme human efforts required to deal with it could be erased with a magnificent equation that would do all the jobs in a much more precise way than this data can convey to us. Therefore, potential equations can emerge from the underlying actions of many interacting factors and tell us how these factors are connected in a deeper fashion.

In chapter 5, we saw how Boltzmann connected the microscopic world to the macroscopic world using a simple but elegant equation. It may be that our observations have not yet yielded an equation, and so we continue to observe and end up in a vicious circle of never-ending data collection. But, once an equation is established, the data becomes obsolete. So, the prediction about the death of equations is unwarranted and, in fact, it represents our inability to connect the data points in a deeper way by using equations.

In chapters 1 and 5, we saw how equations serve as a metaphor to strengthen the theme of this book: humans are transient, but equations are eternal. Such illustrations continue even today,

HERE LIES WHAT WAS MORTAL OF

$$T = \frac{\hbar c^3}{8\pi G M k}$$

STEPHEN HAWKING
1942-2018

The gravestone for Hawking in Westminster Abbey, close to the graves of Isaac Newton and Charles Darwin. Image courtesy of Dean & Chapter of Westminster.

and once again they show us the significance of equations in dealing with the unknown and their inherent strength in adorning the transitory beings we are.

In 2018, when Stephen Hawking died, in a quite emblematic move, his ashes were interred not far from the graves of Isaac Newton and Charles Darwin at Westminster Abbey. His gravestone (see the above diagram) depicts a series of rings surrounding a darker central ellipse. The characters of Hawking's equation express his idea that black holes in the universe are not entirely black but emit a glow, which has become known as Hawking radiation.

By now, you know what the symbols in this equation stand for as we have seen many of these during the course of this book: T – temperature, h – Planck's constant ($\hbar = h/2\pi$), c – the speed of light, G – Newton's gravitational constant, M – mass of the black hole, and k – Boltzmann's constant.

In fact, Hawking's discovery (along with support from his colleague Jacob Bekenstein) opened up new ways of understanding black holes since observation and data collection, in the traditional sense, did not even exist. So, once again, what comes to our rescue is an equation to address one of the most intricate problems in physics.

The equations, by their mere nature, represent deep and subtle relations among everything in our universe, both tangible and intangible. In these ten chapters, we have seen how they span across the disciplines and have the potential to unlock the future while remaining unchanged. The symbols and notations we use to create equations may vary, but the deep desire to connect with everything will never vanish—that makes the equations eternal. Unfortunately, the history of humankind, as noted by Franz Kafka in *The Blue Octavo*

Notebooks, is the instant between two strides taken by a traveler. However, our trivial existence on this planet and the brevity of life should not make us feel frivolous but rather should prepare us to explore further and find the significance of the profound connections nature exhibits in the form of equations that exist in perpetuity.

www.ingramcontent.com/pod-product-compliance
Lightning Source LLC
Chambersburg PA
CBHW072047190526
45165CB00019B/2010